PIPE TEMPLATE LAYOUT

PIPE TEMPLATE LAYOUT

THOMAS W. FRANKLAND
Author of
PIPE TRADES POCKET MANUAL • PIPEFITTING BLUEPRINT READING • SIMPLIFIED SOLUTION OF PIPING OFFSETS • THE PIPE FITTER'S AND PIPE WELDER'S HANDBOOK

McGraw-Hill

New York, New York   Columbus, Ohio   Mission Hills, California   Peoria, Illinois

© 1967 THE BRUCE PUBLISHING COMPANY

Printed in the United States of America

All rights reserved. No part of this book may be reproduced or transmitted in any form or by any means, electronic or mechanical, including photocopying, recording, or by any information storage and retrieval system, without permission in writing from the Publisher.

Send all inquiries to:
GLENCOE/McGraw-Hill
15319 Chatsworth Street
Mission Hills, CA 91345

Library of Congress Catalog Card Number: 67-18213

ISBN 0-02-802400-1

20 21 22 23   99 98 97 96

PREFACE

The author has spent the greater part of his life working in the pipefitting trade. He served his apprenticeship to the trade, worked as a journeyman pipefitter and as a welder and taught the pipefitter apprentices the subjects of welding, drawing, and mathematics for twenty-one years in the Washburne Trade School in Chicago, Illinois.

This book was prepared with the needs of both the apprentice and journeyman to the pipe trades in mind.

The text arrangement is suitable for either individual or classroom study in vocational, trade, and technical high schools where template layout is taught. It is also ideal for home study and on-the-job use.

Technical knowledge of pipe template making constitutes the difference between a good pipe welder and one who can just weld. A good layout man is much preferred over one who lacks this skill and knowledge.

This book explains to the pipe welder how to construct a template by means of clear illustrations and detailed instructions for the various types of fittings that can be fabricated from pipe.

Templates are used for making fittings where accuracy in layout is required. On large size pipe it is more accurate to lay out the cut lines with a template, rather than to use the wrap-around method with no template.

The various drawings have been separated to show clearly each step that is required to produce a template. After the working principles of template making are understood, the views can be superimposed one on the other to save time in drawing.

We feel sure that this book will prove an invaluable aid to the young apprentice as well as to the experienced journeyman in the pipe trades and to anyone else who is interested in pipe fabrication.

CONTENTS

Preface	v
Cutting and Drawing Procedures	1
Miter Turns	9
Angled Plate and Pipe Miter	14
Double Angled Plate and Pipe Miter	19
True Y Fitting	25
Full Size Lateral	30
Reducing Lateral	34
Full Size Tee	38
Reducing Tee — Branch Enters Header	43
Reducing Tee with Branch Pipe Outside Header	48
Eccentric Tee	53
Branch Pipe from the Back of a Welding Elbow	57
Orange Peel Cap	64
Bull Plug Cap	66
Concentric Reducer	71
Eccentric Reducer	73
Decimal Equivalents of Fractions of an Inch	79
Standard Pipe Data	80
Index	81

PIPE TEMPLATE LAYOUT

CUTTING AND DRAWING PROCEDURES

Templates are patterns used to fabricate a welding fitting from pipe. Templates are generally made on drawing paper. If a paper template is used many times, it will deteriorate and become unreliable. It is advisable, therefore, to lay the template out on light-gauge sheet metal.

Care should always be used in developing templates, since the fitting will only be as accurate as the pattern it is made from.

How to Use a Cutting Torch

Cutting the pipe after marking the pattern on it requires some care. When using the cutting torch on a miter cut, the cutting tip must point to the line on the opposite side of the pipe at all times, as shown in Figure 1, so the completed cut will be the

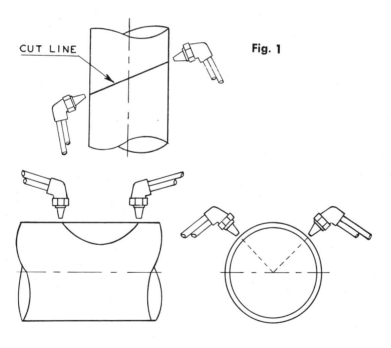

Fig. 1

Fig. 2

same as if it were cut with a hacksaw. The edges are beveled after the cut is made. Never cut and bevel at the same time.

To make an opening for a tee, a radial cut is required. This means that the torch tip is pointed to the center of the pipe at all times as shown in Figure 2. The edges are beveled after the cut is made. The branch pipe is also radial cut.

Drawing Equipment

The drawing equipment needed to make a template is as follows:

1 — Drawing Board
1 — T-Square
1 — 45-deg. Triangle
1 — 30 by 60-deg. Triangle
1 — Ruler

1 — Irregular Curve
     (French Curve)
1 — Mechanical Drawing Set
1 — Pencil Eraser
1 — Drawing Pencil
     Drawing Paper

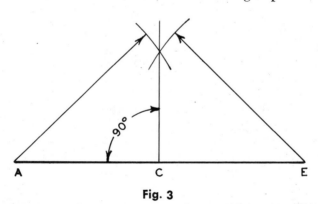

Fig. 3

How to Bisect a Line

Figure 3 shows the line A–E bisected, or divided into two equal parts. Set the compass on A and open it up to any setting past the center between A and E and scribe an arc. With the same compass setting, set the compass on E and scribe another arc so it intersects the first arc. Where the arcs cross will be the exact center. At right angles (90-deg. angle) to the line A–E, draw a line from the intersection of the arc lines to the A–E line, locating point C, the center between A and E.

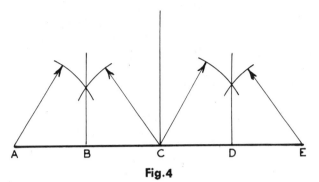

Fig. 4

How to Divide a Line Into Four Equal Parts

Figure 4 shows the line A–E from Figure 3 divided up into four equal parts. Bisect the space from A to C, by setting the compass on A and opening it up to any setting past the center between A and C and scribe an arc. With the same compass setting, scribe an arc from point C, then swing the compass over from C and scribe an arc between C and E. Then set the compass on point E and scribe another arc. Where the arcs intersect, drop a line down to the A–E line locating points B and D. This will divide the line into four equal parts.

How to Divide a Line Into Six Equal Parts

Figure 5 shows a line divided into six equal parts. Set the compass anywhere past the center between points A and G and scribe an arc with the compass set on point A. With the same compass setting, scribe an arc with the compass set on point G. Draw a line

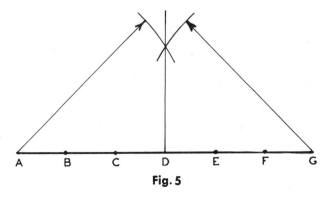

Fig. 5

from the intersection of the arc lines to line A–G, locating point D. This line must be at right angles to line A–G. Line A–G is now divided into two equal parts.

Next divide the space A–D and D–G, each into three equal parts by using a pair of dividers. By trial and error, set the dividers for ⅓ of the distance between A and D locating points B and C. With the same setting of the dividers, divide the space from D to G into three equal parts, locating points E and F. The line A–G is now divided into six equal parts.

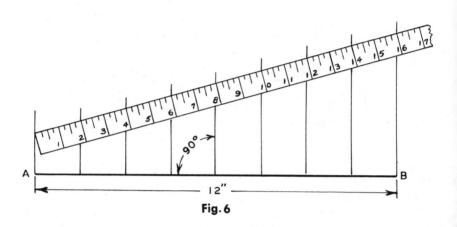

Fig. 6

How to Divide a Line Into Equal Parts With a Rule

Lay a rule on a diagonal between the vertical lines A and B (Fig. 6). Use the smallest scale which has the number of spaces required and that is greater in length than the space to be divided.

If line A–B is 12 in. long and is to be divided up into eight equal parts, lay a rule diagonally across the vertical A and B lines with the end of the rule touching line A and the 16-in. mark touching line B. Project the 2-, 4-, 6-, 8-, 10-, 12-, and 14-in. divisions on the rule down to the line A–B, laying out the required eight divisions. All vertical lines to be drawn at right angles with line A–B.

How to Bisect an Angle

The point where the lines meet to form the angle will be the compass point. Set the compass with any radius setting, and scribe

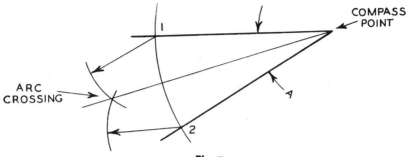

Fig. 7

an arc intersecting the straight lines. Figure 7 shows the straight line and arc intersections numbered 1 and 2.

Set the compass on point 1 and open it up to any setting past the center between points 1 and 2, and scribe an arc. The compass setting must pass the center, so the arcs will cross.

With the same compass setting, place the compass on point 2 and scribe another arc. Where the arcs cross will be the center. With a straightedge, line up the arc intersection and the compass point and draw a straight line bisecting angle A into two equal parts.

How to Divide a Quarter Circle Into Eight Equal Parts

Line up the 45-deg. triangle with the compass point and draw a line to the arc, locating point 5, as shown in Figure 8–A.

With the compass, bisect the angles between points 1 and 5, and 5 and 9, locating points 3 and 7 as shown in Figure 8–B.

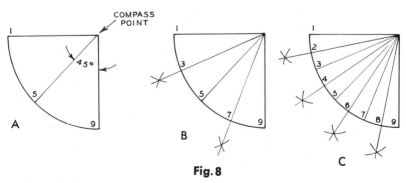

Fig. 8

With the compass, bisect the angles between 1 and 3, 3 and 5, 5 and 7, and 7 and 9, locating points 2, 4, 6, and 8 as shown in Figure 8–C. The quarter circle is now divided into eight equal parts.

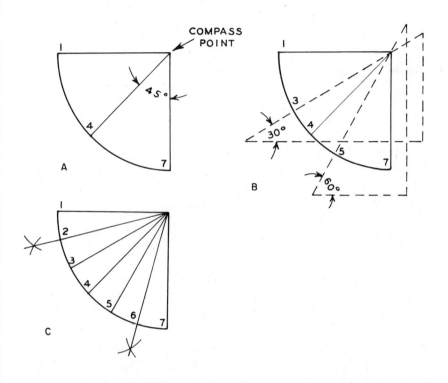

Fig. 9

How to Divide a Quarter Circle Into Six Equal Parts

To divide a quarter circle into six equal parts, first line up the 45-deg. triangle with the compass point, then draw a line to the arc locating point 4 as shown in Figure 9–A.

Next use the 30 by 60 deg. triangle and draw lines from the compass point to the arc, locating points 3 and 5 as shown in Figure 9–B.

Then with the compass, bisect the angles between points 1 and 3, and 5 and 7, locating points 2 and 6 as shown in Figure 9–C. Or set the dividers for the distance between points 4 and 5, and lay out this measurement between 1 and 3, and 5 and 7, locating points 2 and 6.

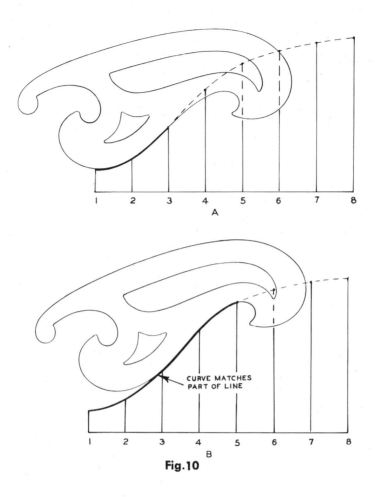

Fig. 10

How to Use the Irregular Curve

RULE: The irregular curve must connect a minimum of three points.

Figure 10-A shows the start of laying out the curved line with a part of the irregular curve connecting the points on ordinate lines 1, 2, and 3. This part of the irregular curve is found by moving it around until the proper curve is located to join these three points.

With a pencil connect these three points with a curved line, using the edge of the irregular curve as a guide.

If the irregular curve connects four or five points, this is all the better. But no less than three points can be used.

Once the curve has been established and until it is completed, the following procedure must be followed. The edge of the irregular curve must again line up three or more points. Figure 10–B shows one point that has already been used on ordinate line 3, and two new points on ordinate lines 4 and 5 lined up with the edge. Part of the irregular curve must also line up with a short part of the line already established. This will eliminate any humps at the points, or depressions between the points. When the curved line is completed, it must be a smooth-flowing line with no humps or depressions. Connect points 3, 4, and 5 with a pencil line. Continue with this method until the curved line is completed.

There are variously-shaped irregular curves. It is best to have two or three differently-shaped curves on hand. If you cannot find the proper curve required on one, you may find it on another.

MITER TURNS

To make a layout for a miter turn, draw a base line equal in length to the outside diameter of the pipe, plus the outside circumference, plus ½ in. for spacing between.

Lay out a semicircle with a diameter equal to the outside diameter of the pipe, at a distance above the base of about two times the outside diameter of the pipe. Divide this semicircle into eight equal parts. Number these points from 1 to 9 as shown in Figure 11–A. Project each point down to the base line. The semicircle could be laid out below the base line if desired.

At the intersection of line 9 and the base line, lay out the cut line for the angle of cut desired. This will establish a view of the pipe after it has been cut. To determine the angle of cut, use the following formula:

Angle of cut = total angle of fitting ÷ 2 times the number of welds.

Example: If the angle of the fitting is 45° and has one weld, applying the above formula:

1 weld × 2 = 45° ÷ 2 = 22½ angle of cut.

On the base line, ½ in. from this view, lay out the outside circumference of the pipe as shown in Figure 11–B. Divide this length into sixteen equal parts (always twice as many parts as there are on the semicircle). Number each point from 1 to 9 as shown. Project lines vertically from each point at convenient lengths. These lines are called ordinate lines.

Point 9 will be located on the base line of the template. The point at which line 8 from the semicircle intersects the cut line will establish the length of ordinate 8 line on the template. Project the intersection of these two lines from Fig. 11–A to the left to ordinate lines 8, where the lines intersect on the template, and mark the length with a dot. Project the intersection of line 7 from the semicircle and the cut line to the left to ordinate line number 7 and mark the intersections with a dot. Repeat this with lines 6, 5, 4, 3, 2, and 1.

Connect these dots on the template with a curved line by using an irregular curve. When using an irregular curve always remember

9

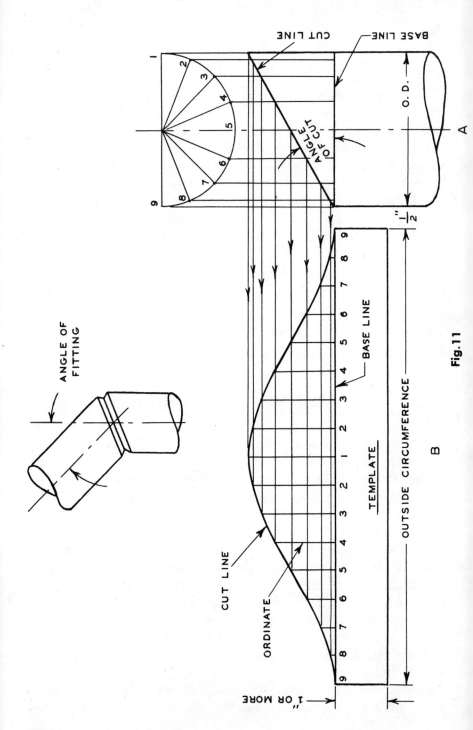

Fig.11

to line up at least three points at one time as explained in Figure 10-A and 10-B.

Add at least 1 in. below the base line on the template to provide a margin to be used as a wrap-around to aid in lining up the template on the pipe.

By this method, miter-turn templates can be laid out for any size of pipe, or any angle of fitting. The number of divisions of the semicircle and of the base line on the template depends upon the size of pipe to be used. Larger pipes require more divisions. The more divisions used the more accurate the template.

Cut the end of the pipe on a miter, then bevel. Never cut and bevel at the same time. See Figure 1 on how to make a miter cut.

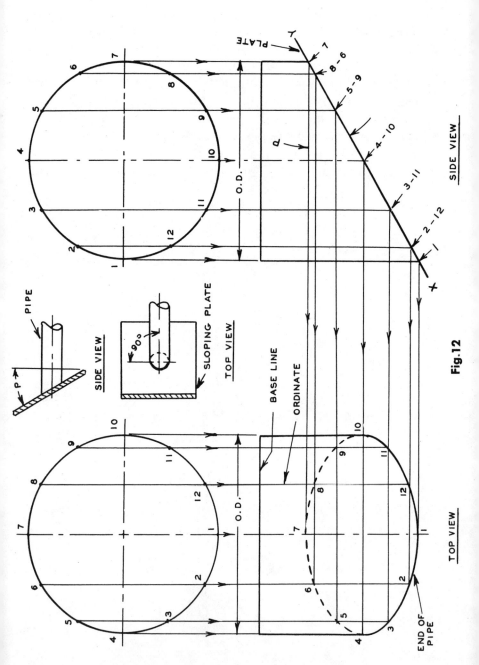

Fig.12

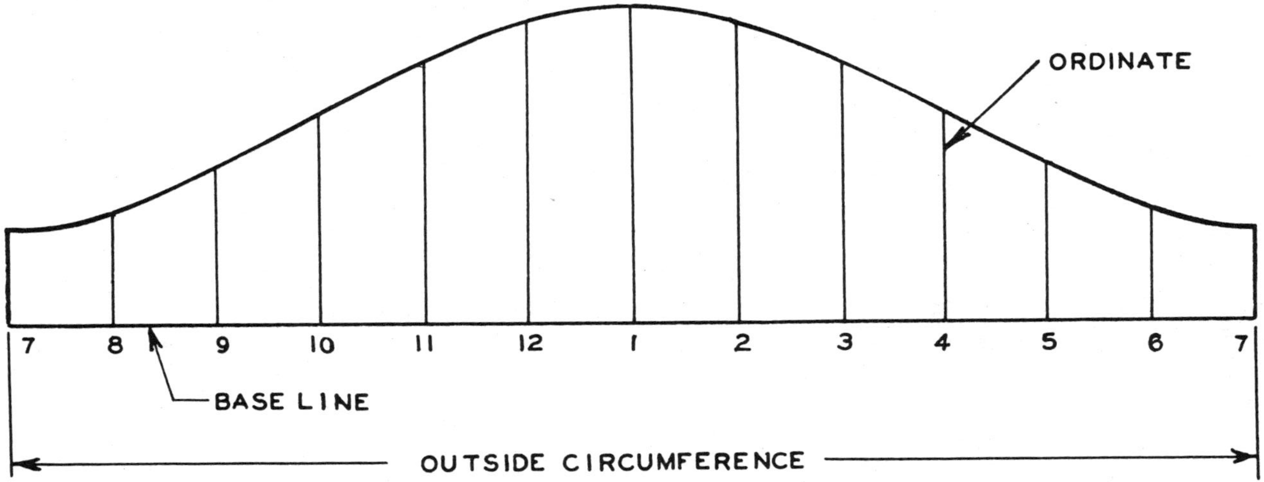

Fig. 13

BRANCH TEMPLATE

ANGLED PLATE AND PIPE MITER

On the top view of Figure 12 lay out a vertical center line; then lay out a full circle with a diameter equal to the outside diameter of the pipe. Divide the circle into twelve equal parts. Number each division from 1 to 12 as shown. Parallel to the center line, project each division down at a random length.

On the side view draw a vertical center line, and scribe another circle with the same diameter as the top view and divide it into twelve equal parts. Number each division from 1 to 12 as shown, and project each division down at a convenient length parallel with the center line.

A short distance below the circle draw the line $X-Y$ (representing the plate) on the required angle P. Number the intersections where the vertical lines from the circle intersect the $X-Y$ line as shown, 1, 2-12, 3-11, etc.

From the $X-Y$ line on the side view project point 1 to the left to the top view, and mark it with a dot where it intersects line 1.

Project points 2 and 12 from the side view to the corresponding lines 2 and 12 on the top view and mark the intersection of the number 2 lines with a dot; also mark the intersection of the number 12 lines with a dot.

Repeat this with points 3-11, 4-10, 5-9, 6-8, and 7. Connect these points on the top view with a curved line, using an irregular curve to form the line. The curve establishes the end of the pipe.

How to Lay Out the Branch Template

Figure 13 illustrates the template. Lay out a base line equal to the outside circumference of the pipe. Divide this base line into twelve equal parts, the same number of parts as used on the circle. Number these points from 1 to 12 as shown. Project an ordinate line up from each point at random lengths. On the top view of Figure 12 lay out a base line anywhere between the circle and the oval that represents the end of the pipe.

With a pair of dividers or a compass, take the length of ordinate 1 from the base line to the oval and transfer this length to ordinate 1, measuring up from the base line as shown in Figure 13. Separately take each of the other ordinate lengths from 2 to 12 from the base line to the end of the pipe in Figure 12, and transfer their lengths to their corresponding ordinate lines on the template in Figure 13. Connect these points with a curved line to complete the branch template.

How to Lay Out the Opening Template

The author has separated Figures 12 and 14 for clarity, but for practical purposes one could be superimposed on the other (combining the two drawings) to save layout time.

On the top view of Figure 14 lay out a vertical center line; then lay out a full circle with a diameter equal to the inside diameter of the pipe. Divide the circle into twelve equal parts. Letter each division from A to L as shown. Project each division down at a convenient length parallel with the center line.

On the side view draw a vertical center line, and scribe another circle with the same diameter as the top view circle; then divide the circle into twelve equal parts. Both circles must have the same number of divisions. Letter each division from A to L as shown; and project each division down parallel to the center line at a convenient length.

At any convenient distance below the circle, draw the X–Y line which represents the face of the plate, on the required angle P. Letter the intersections where the vertical lines from the circle intersect the X–Y line as shown at A, B–L, C–K, etc.

From the X–Y line on the side view, project point A to the left to the top view and mark it with a dot where it intersects line A. Project points B and L from the side view to the corresponding lines B and L on the top view and mark the intersection of the B lines with a dot; also mark the intersection of the L lines with a dot.

Repeat this with points C–K, D–J, E–I, F–H, and G. Connect these points on the top view with a curved line, using an irregular curve to aid you in forming the line. The curve establishes the opening.

Figure 15 illustrates the opening template. Lay out a base line at a length longer than the distance from A to G on the X–Y line of the side view of Figure 14.

Locate the center of this base line and letter this point D–J.

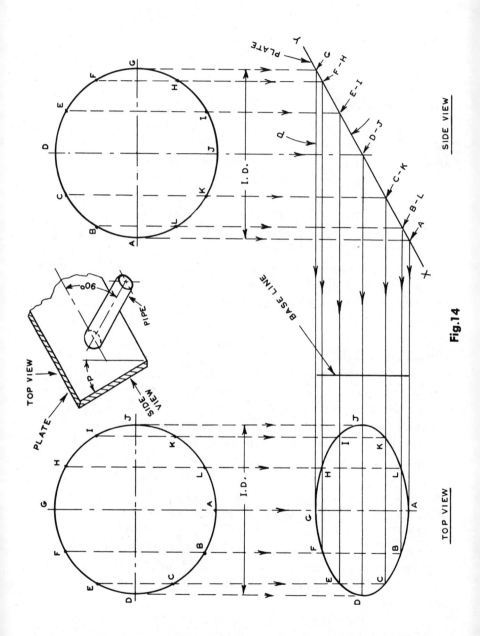

Fig.14

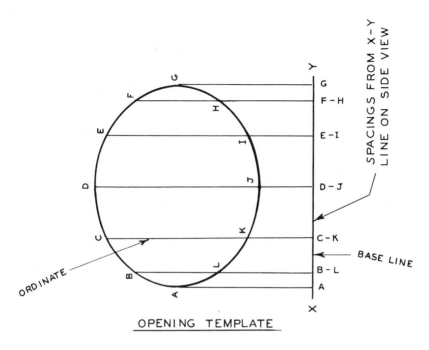

Fig. 15

Extend an ordinate line from this point at a random length.

With a pair of dividers or a compass, from the X–Y line in Figure 14 take the length of the spacing from point D–J to point C–K, and lay out this measurement from point D–J on the base line in Figure 15. Mark this point with the letters C–K as shown, and extend an ordinate line of indefinite length.

Take the space length from point C–K to point B–L on the X–Y line in Figure 14 and transfer it to the base line on Figure 15 as shown; then extend an ordinate line of indefinite length.

Take the space length from point B–L to point A on the X–Y line of Figure 14, and transfer it to the base line in Figure 15 as shown, then extend an ordinate line of indefinite length.

Transfer the other spaces D–J to E–I, E–I to F–H, and F–H to G from the X–Y line in Figure 14 to the base line in Figure 15, and extend ordinate lines from the base for each.

On Figure 14 draw a base line anywhere between the side view and the oval on the top view as shown; then with a pair of dividers or a compass take the length from the base line to point A on the oval and lay out this length on the ordinate line *A* from the base line in Figure 15.

Take the length from the base line to the point *B* on the oval in Figure 14 and lay out this length from the base line on the ordinate line *B–L* in Figure 15, locating point *B*.

Take the length from the base line to the point *L* on the oval in Figure 14 and lay out this length from the base line on the ordinate line *B–L* in Figure 15, locating point *L*.

Repeat this with all the other points, *C, K, D, J*, etc. Connect all the points in Figure 15 with a curved line, using an irregular curve, to complete the opening template.

The end of the pipe will be miter cut as shown in Figure 1, then beveled, and will fit up against the face of the plate; it will not enter the opening.

The hole will be cut with the torch tip held parallel with the pipe center line. Do not bevel the opening.

The number of divisions of the circles and of the base line on the templates depends upon the size of the pipe used. Larger pipes require more divisions. The more divisions used the more accurate the templates. The number of divisions on the full circles and on the base line must always be the same. For example: if there are sixteen divisions on the circle, there must be sixteen divisions on the base line of the pipe template.

DOUBLE ANGLED PLATE AND PIPE MITER

On the top view of Figure 16, lay out the center line for the number of degrees of angle N; then lay out a full circle with a diameter of the outside diameter of the pipe.

Divide the circle into twelve equal parts. Number each division from 1 to 12 as shown. Parallel to the center line, project each division down at a convenient length.

On the side view draw a vertical center line, and scribe another circle with the same diameter as on the top view. Also divide the circle into twelve equal parts. Number each division from 1 to 12 as shown, and project each division down parallel with the center line.

A short distance below the circle draw the line $X-Y$ which will represent the face of the plate on the required angle M. Number the intersections where the vertical lines from the circle intersect the $X-Y$ line as shown, 1, 2-12, 3-11, etc.

Project point 7 from the $X-Y$ line on the side view to the left to the top view and mark it with a dot where it intersects line 7. Project points 6 and 8 from the side view to the corresponding lines 6 and 8 on the top view, and mark the intersections with a dot. Repeat this with the 5-9, 4-10, 3-11, 2-12, and 1 points. Connect these points on the top view with a curved line, using an irregular curve (French curve). This curve establishes the end of the pipe.

How to Lay Out the Branch Template

Figure 17 illustrates the template. Lay out a base line equal to the outside circumference of the pipe. Divide this base line into twelve equal parts, the same number of parts as in the circle. Number these points from 1 to 12 as shown. Project an ordinate line up from each point. On the top view of Figure 16 lay out a base line anywhere between the circle and the oval that represents the end of the pipe.

Separately measure the ordinate length of points 1 to 12 from the base line to the end of the branch line in Figure 16, and transfer each measurement to the respective ordinate lines on the template

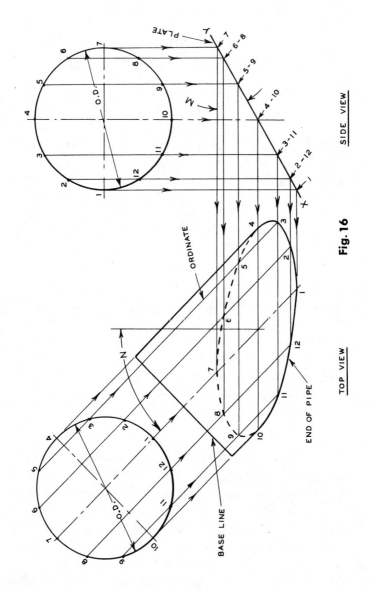

Fig. 16

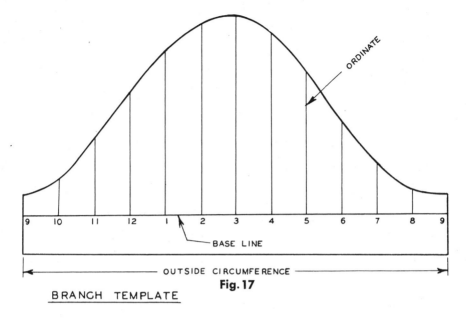

Fig. 17

BRANCH TEMPLATE

on Figure 17. Connect these points using an irregular curve (French curve) to complete the branch template.

How to Lay Out the Opening Template

The author has separated Figures 16 and 18 for clarity. The two views could be combined, or superimposed, one over the other to save time in layout.

On the top view of Figure 18 lay out the center line for angle *N;* then lay out a full circle with a diameter equal to the inside diameter of the pipe. Divide the circle into twelve equal parts. Letter each division from *A* to *L* as shown. Project each division down at a random length parallel to the center line.

On the side view draw a vertical center line, and scribe another circle with the same diameter as the top view circle. Divide this circle into twelve equal parts. Letter each division from *A* to *L* as shown. Parallel to the center line, project each division down.

Below the circle at any convenient distance draw the *X–Y* line on the required angle *M*. Letter the intersections where the vertical lines from the circle intersect the *X–Y* line as shown, *A*, *B–L*, *C–K*, etc.

21

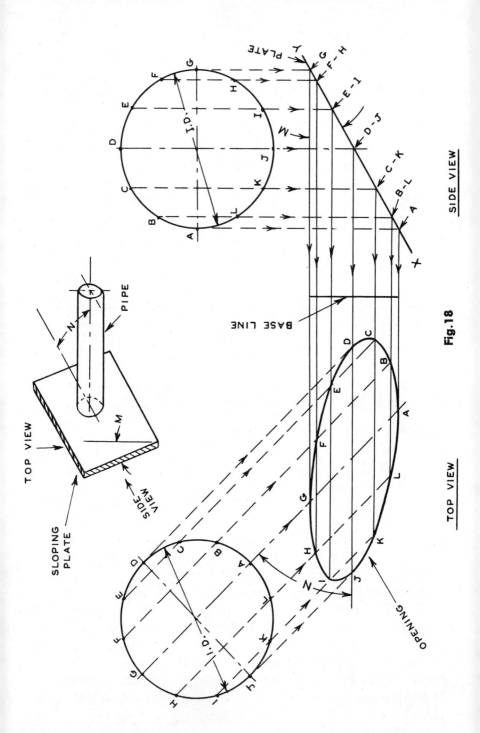

Fig.18

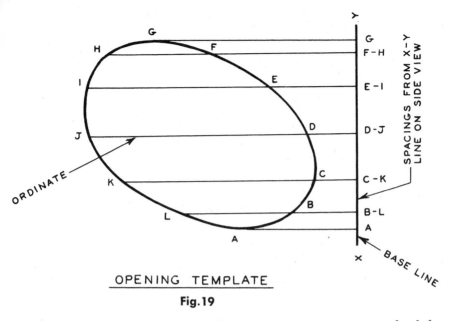

OPENING TEMPLATE

Fig. 19

From the X–Y line on the side view, project point A to the left to the top view and mark it with a dot where it intersects line A.

Project points B and L from the side view to the corresponding lines B and L on the top view. Mark the intersection of the B lines with a dot; also mark the intersection of the L lines with a dot.

Repeat this procedure with points C–K, D–J, E–I, F–H, and G. Connect these points on the top view with a curved line, using an irregular curve to aid you in forming the line. The curve establishes the opening.

Figure 19 illustrates the opening template. Lay out a base line at a length equal to the distance from A to G on the X–Y line on the side view of Figure 18.

Locate the center of this base line and letter this point D–J. Extend an ordinate line from this point to an indefinite length. From line X–Y in Figure 18, with a pair of dividers or a compass, take the spacing from point D–J to point C–K, and lay this spacing out from point D–J on the base line in Figure 19. Mark this point with the letters C–K as shown, and extend an ordinate line of convenient length.

Measure the distance from point C–K to point B–L on the X–Y line in Figure 18 and transfer this measurement to the base line on Figure 19. Lay this spacing out from point C–K. Mark this point

23

with the letters *B–L* as shown, and extend an ordinate line of indefinite length.

Measure the distance from point *B–L* to point *A* on the *X–Y* line in Figure 18, and transfer this measurement to the base line in Figure 19. Lay this space out from point *B–L*. Mark this point with the letter *A* as shown, and extend an ordinate line of indefinite length.

Transfer the other spaces *D–J* to *E–I*, *E–I* to *F–H*, and *F–H* to *G* from the *X–Y* line in Figure 18 to the base line in Figure 19, and extend ordinate lines from the base line.

On Figure 18 draw a base line anywhere between the side view and the oval on the top view as shown. With a pair of dividers or a compass take the length from the base line to the *A* point on the oval, and lay out this length on ordinate line *A*, from the base line in Figure 19.

Take the length from the base line to the point *B* on the oval in Figure 18 and lay out this length from the base line on ordinate line *B–L* in Figure 19, locating point *B*.

Take the length from the base line to point *L* on the oval in Figure 18 and lay out this length from the base line on the ordinate line *B–L* in Figure 19, locating point *L*. Repeat this with all the other points, *C*, *K*, *D*, *J*, etc.

To complete the opening template, connect all points in Figure 19 with a curved line, using an irregular curve (French curve).

The end of the pipe is miter cut, as explained in Figure 1, then beveled. It will fit up against the face of the plate, and will not enter the hole. The hole will be cut with the torch tip held parallel with the pipe center line, at angle *N* to the plate. If angle *N* is 45 deg., then the cutting tip will be held on a 45-deg. angle when cutting the opening. Do not bevel the opening.

The number of divisions of the circles and of the base lines on the templates depends upon the size of the pipe used. Larger pipes require more divisions. The larger the number of divisions used the more accurate the template will be. The number of divisions on the full circles and on the base lines of the templates must always correspond. If there are sixteen divisions on the circle, there must be sixteen divisions on the base line of the template for the pipe.

TRUE Y FITTING

To make a true Y fitting template, lay out the branch center lines on whatever angle is required. Add the main run center line as shown in Figure 20. Lay out one half of the outside diameter of the pipe on each side of the center lines. Draw these outside lines parallel with the center lines until they intersect at points A, B, and C. Connect these points with the intersection of the center lines at D, establishing the ends of the three pipes.

Measure back an inch from the intersection of the branch lines at point C, and draw the base line 1–9. Set the compass at the intersection of this base line and the center line, and draw a semicircle with a diameter equal to the outside diameter of the branch pipe. Divide this semicircle into eight equal parts. Number these points from 1 to 9 as shown. Project each point to the end of the branch pipe. Each line is to be drawn parallel with the center line.

How to Lay Out the Branch Template

Lay out a base line equal to the outside circumference of the branch pipe as shown in Figure 21. Divide this line into 16 equal parts. Number these points on the base line from 1 to 9, and 9 to 1, as shown. Draw ordinate lines from these points at right angles to the base line.

Measure the length of ordinate number 1 in Figure 20, from the base line to the end of the branch pipe, and transfer this length to the number 1 ordinate lines on the template, as shown in Figure 21. Transfer the length of ordinate number 2, from the base line to the end of the pipe in Figure 20 to the number 2 ordinate lines on the template. Repeat this for the ordinate lengths for ordinates number 3, 4, 5, 6, 7, 8, and 9. Connect the ends of these ordinate lines with a smooth curved line, using an irregular curve (French curve), to complete the template. This template will be for both branch pipes since they are on the same angle. If the angles were different, then two templates would have to be developed.

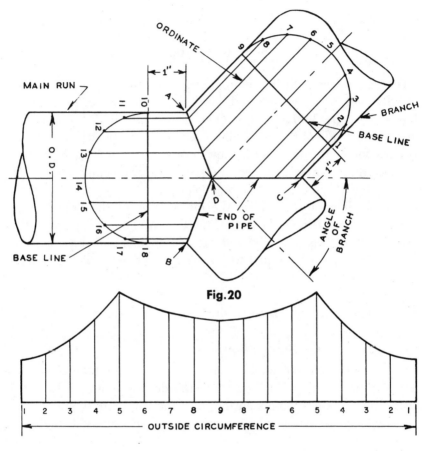

Fig. 20

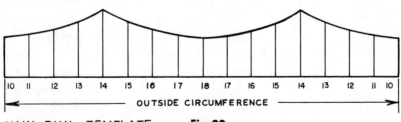

BRANCH TEMPLATE   **Fig. 21**

MAIN RUN TEMPLATE   **Fig. 22**

How to Lay Out the Main Line Template

In Figure 20 measure an inch from the intersection of the branch and main run lines at points A and B and draw the base line 10–18. Set the compass at the intersection of the base line and the center line, and draw a semicircle with a diameter equal to the outside diameter of the pipe. Divide this semicircle into eight equal parts. Number these points from 10 to 18 as shown. Project each point to the end of the pipe. Each line must be parallel with the center line.

Lay out a base line with a length equal to the outside circumference of the main pipe, as shown in Figure 22. Divide this line into sixteen equal parts. Number these points from 10 to 18, and 18 to 10 as shown on the base line. Draw ordinate lines at random lengths, at right angles to the base line from each point.

Measure the length of ordinate line number 10 in Figure 20, from the base line to the end of the pipe, and transfer this measurement to the number 10 ordinate lines on the template (Fig. 22). Measure the length of ordinate line number 11 in Figure 20, and transfer this measurement to the number 11 ordinate lines on the template. Repeat this for the ordinate lengths 12, 13, 14, 15, 16, 17, and 18. Connect the ends of these ordinates with a smooth-curved line to complete the template.

Make a miter cut with the cutting torch. As you cut the ends of the pipes, imagine you are making the cuts with a hacksaw. Bevel the ends.

On larger-size pipes, more divisions would be used for accuracy. The layout procedure would be exactly the same.

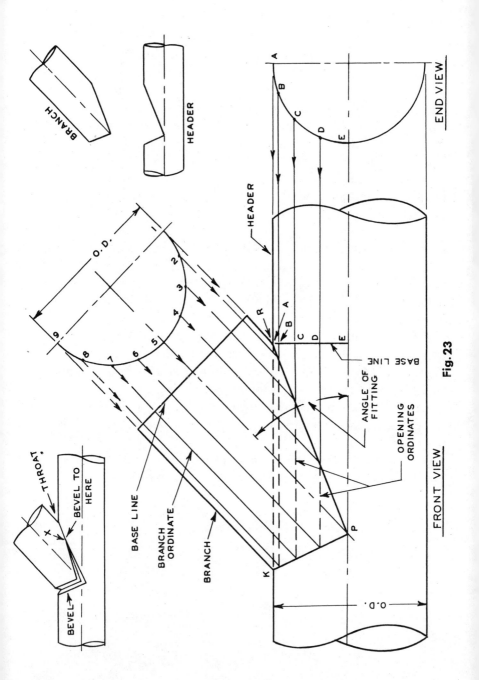

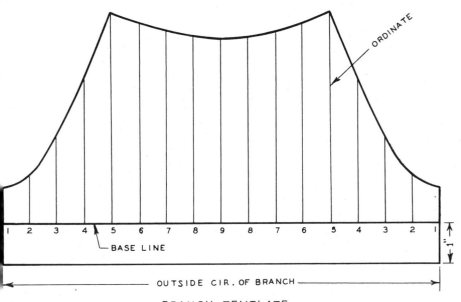

BRANCH TEMPLATE

Fig. 24

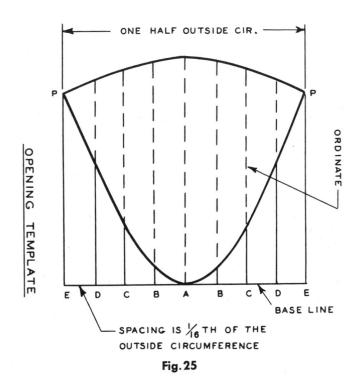

SPACING IS 1/16 TH OF THE OUTSIDE CIRCUMFERENCE

Fig. 25

FULL SIZE LATERAL

To make a template for a full size lateral, lay out a horizontal center line on the front view for the header pipe. Then lay out a center line for the branch pipe on the angle required for the fitting. Next lay out the outside diameter of each pipe half on each side of the center line. Then draw the lines representing the outside of each pipe as shown in Figure 23.

With a straightedge, connect point P with points K and R. These lines represent the end of the branch pipe and the edge of the opening.

On the branch center line scribe a semicircle with a diameter equal to the outside diameter of the pipe, and divide the semicircle into eight equal parts. Number each division from 1 to 9 as shown.

Parallel to the center line, project each division down to the end of the pipe. Anywhere above point R draw a base line.

How to Lay Out the Branch Template

In Figure 24 lay out a base line equal to the outside circumference of the pipe. Divide this line into sixteen equal parts, i.e., always twice as many parts as used on the semicircle. Number each division from 1 to 9 as shown. Extend an ordinate line up from each division.

With the dividers or a compass, measure the length of ordinate 1 in the front view of Figure 23, from the base line to point R, and transfer this length to the ordinate 1 lines in Figure 24, marking each point with a dot.

Measure the length of ordinate 2 (Fig. 23), from the base line to the end of the pipe, and transfer this length to the number 2 ordinates in Figure 24, marking each point with a dot. Repeat this with ordinates 3, 4, 5, 6, 7, 8, and 9.

Using an irregular curve, connect each point with a curved line. Below the base line measure down a margin of an inch or more to aid in lining up the template on the pipe when laying out the cut line.

How to Lay Out the Opening Template

On the end view of Figure 23, scribe a semicircle with a diameter equal to the outside diameter of the pipe. Divide the upper quarter circle into four equal parts. Letter each point from A to E. Project each division to the K–P line on the front view.

Lay out a vertical base line from point R to the center line.

In Figure 25 lay out a base line equal in length to one half the outside circumference of the pipe. The spacing between each ordinate line is $\frac{1}{16}$ of the outside circumference. Divide the circumference by sixteen to obtain each spacing length. Lay out the eight divisions and letter each point from A to E as shown. Extend an ordinate line from each division.

From the front view in Figure 23 obtain the length of ordinate A. This length is taken from the base line to point K, or the distance from R to K. Lay out this length on ordinate line A from the base (Fig. 25), and mark it with a dot.

On ordinate line B on the front view, obtain the length from the base line to the P–R line, and lay out this length on the B ordinate lines on the opening template. Mark each with a dot.

On the ordinate line B on the front view, obtain the length from the base line to the K–P line, and lay out this length on the ordinate B lines from the base line on the template. Mark each point with a dot. Repeat this with the C, D, and E ordinate lines. Connect the points with a curved line, using an irregular curve (French curve), to complete the opening template.

Make a miter cut, using the cutting torch, on the end of the branch pipe and the hole. Imagine you are making the cut with a hack saw. After the miter cut, bevel the branch and the hole to the X point on each side of the pipe. See the sketch in Figure 23. The throat is not beveled since it is difficult to reach in and weld to the inner wall. The outer wall from point X through the throat to point X on the other side will form a V for welding.

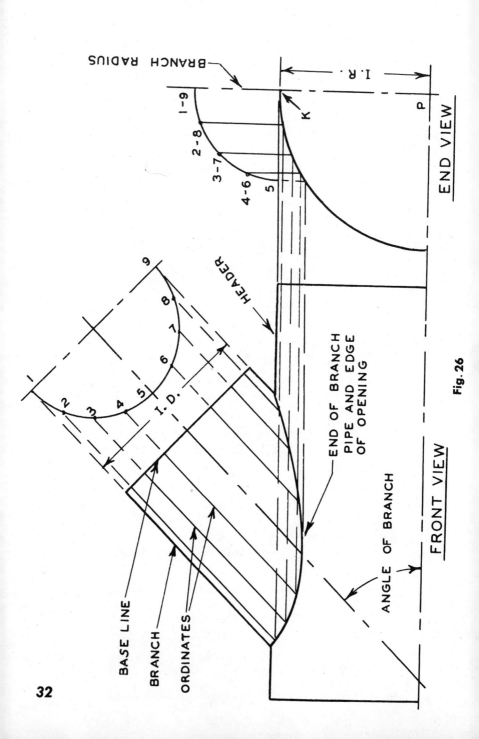

Fig. 26

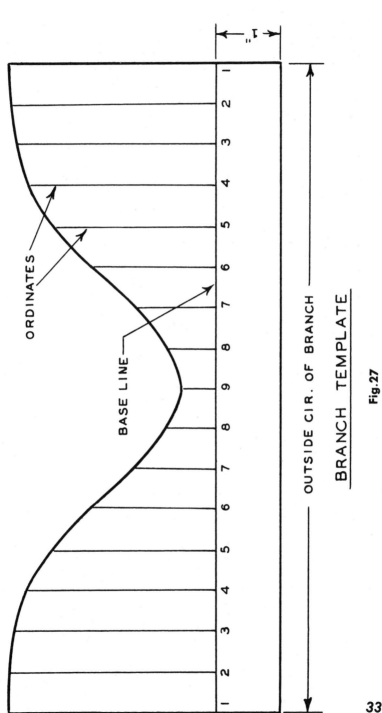

Fig. 27
BRANCH TEMPLATE

REDUCING LATERAL

Inside diameters are used on the layout of the front and end views of the template for a reducing lateral so that the inside walls of the pipe may be fitted for accuracy.

Figure 26 shows the front and end views. Lay out the center lines on the front view for the branch and header as shown. The branch center line is laid out on whatever angle of branch is required.

Scribe a semicircle on the branch center line with a diameter equal to the inside diameter of the branch pipe. Measure up from the header center line, half the inside diameter of the header, and draw a line parallel with the center line. Project the sides of the branch semicircle down to meet this line. The bottom half of the header pipe is not used, so to save space it is not shown.

Divide the semicircle on the branch into eight equal parts. Number each division from 1 to 9 as shown. Project points 2 and 8 down at random lengths parallel to the center line.

On the end view of Figure 26, set the compass at point P, and scribe a quarter circle with a radius equal to the inside radius of the header pipe. At point K set the compass with a setting equal to the radius of the branch and scribe a quarter circle. Divide this quarter circle into four equal parts. Number these points as shown. Project points 2 and 8 on the quarter circle down until it intersects the header quarter circle. Project this point to the left to the front view until it intersects both lines 2 and 8 from the semicircle on the branch, and place a dot at each intersection. Repeat this with points 3–7, 4–6, and 5.

On the front view connect these dots using an irregular curve to lay out the curved line. This curve establishes the end of the branch pipe and also the edge of the opening.

Lay out a base line on the branch pipe, anywhere between the top of the header pipe and the semicircle on the front view.

How to Lay Out the Branch Template

Figure 27 illustrates the branch template. Lay out a base-line length equal to the outside circumference of the branch pipe.

Divide this base line into sixteen equal parts, twice the number of parts of the branch semicircle. Number each point as shown, from 1 to 9. Project ordinate lines up from each point at a random length. In the front view in Figure 26 measure each of the ordinate lengths 1, 2, 3, 4, 5, 6, 7, 8, and 9, from the base line to the curved line (end of branch line). Transfer these lengths to the respective ordinates on the branch template as shown in Figure 27. Connect these points with a curved line to complete the template. Allow a margin of an inch or more below the base line for line-up purposes when the template is wrapped around the pipe.

How to Lay Out the Template for the Opening

Figure 28 will not be drawn as a separate drawing but it will be superimposed on Figure 26. The author has shown it here as a separate drawing for clarity.

In Figure 28, to the left of the opening, lay out a quarter circle with a radius equal to the inside radius of the header. Divide this quarter circle into eight equal parts. Number these points from 1 to 9 as shown.

Draw a base line vertically from the intersection of the branch and the header as shown (point K). Mark the lowest point of the opening on the branch center line as X. Project this point to the right to the base line. Also project this point to the left to the quarter circle and mark this point as X. This point meets the quarter circle between points 4 and 5. As the branch pipe increases, point X will vary and may fall between points 5 and 6 or 6 and 7, depending upon the size of pipe used. Project points 1, 2, 3, and 4 to the right across the opening to the base line.

Figure 29 shows the template for the opening. Lay out the base line as shown at a convenient length. Starting at point 1, lay out points 2, 3, 4, and X each way. These spacings are obtained from the spacings between points 1, 2, 3, 4, and X from the quarter circle in Figure 28. To obtain these curved lengths between the numbers accurately, divide the inside circumference by 32. With a pair of dividers, take the length from 4 to X off the quarter circle, and lay out this measurement on the ends of the base line from points 4. These curved lengths could also be obtained by bending a stiff piece of paper to the curvature of the quarter circle and marking each point on the paper, then transferring these

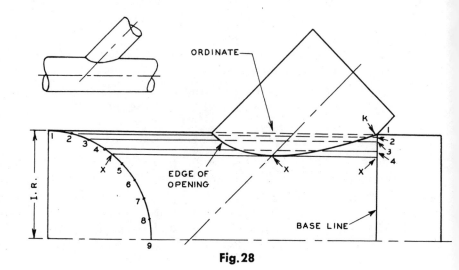

Fig. 28

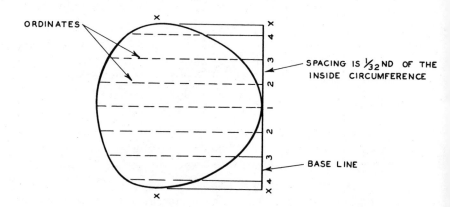

TEMPLATE FOR OPENING

Fig. 29

measurements to the template base line. Draw the ordinate lines out from the base line at convenient lengths.

To lay out the lengths of the ordinates on the opening template, measure the length of ordinate 1 from the base line to the far edge of the opening in Figure 28, and transfer this length to ordinate 1 on the template (Fig. 29).

Measure the dimensions on ordinate 2, from the base line to the near edge and from the base to the far edge of the opening in Figure 28, and transfer these two dimensions to the number 2 ordinates on the template (Fig. 29). Repeat this for ordinates 3 and 4.

To obtain the lengths of ordinates X, measure this length from the base line to point X (lowest point on the branch center line) in Figure 28, and transfer this dimension to the X ordinate lines on the opening template (Fig. 29). Connect these points with a curved line, using the irregular curve to do so, to complete the opening template.

Radial-cut the branch pipe and the opening with a cutting torch; then bevel the opening and the branch pipe.

FULL SIZE TEE

To make a template for a full size tee, lay out the two semicircles for the header on the end view of Figure 30, equal to the inside and outside diameters of the pipe. Above the header lay out a quarter circle with a radius equal to half the outside diameter of the branch pipe. Divide this quarter circle into six equal parts. Letter these points from A to G as shown.

Project the outside diameter of the header to the left, laying out the width of the header pipe for the front view. Draw a vertical center line, and lay out half the outside diameter of the branch on each side. Where these lines intersect the top of the header (at X), connect these points with a 45-degree angle to the intersection of the horizontal and vertical center lines. This point is rounded off to increase the strength of the fitting.

To round off this point, measure up from the center line on the outer semicircle of the header on the end view, the distance of two times T to locate point P. (T is the pipe wall thickness.) Project point P on the end view to the branch center line on the front view. Locate a compass point on the branch center line by trial and error; then scribe an arc connecting the two 45-deg. lines and point P to round off this point with a smooth curve.

On the end view, line up point P with the center of the header with a straightedge, and draw a line from P to the inside wall locating point K. Measure up one T dimension from point K, locating the end of the branch pipe. At this point draw a short horizontal line to the left. Project point A down from the quarter circle above, to this horizontal line. Where they intersect locates point M. Set the compass for half the outside diameter of the header and scribe an arc, using point M as a compass point. Scribe another arc with the same compass setting, using point L as a compass point. Where these two arcs intersect will locate point R.

Set the compass on point R, with a compass setting of half the outside diameter of the header, which is shown as radius S; scribe an arc line from L to M. This arc will establish the end of the branch pipe.

Draw a base line about an inch above point L. Project points B, C, D, E, F, and G down from the quarter circle to the end of

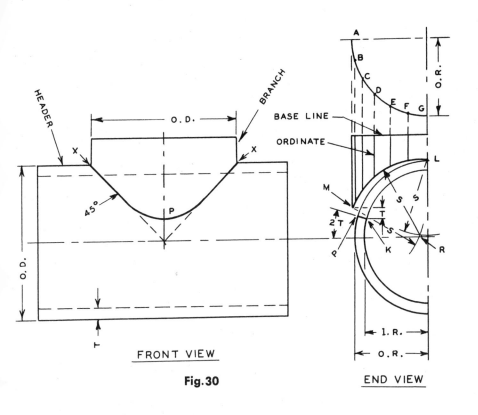

Fig. 30

FRONT VIEW · END VIEW

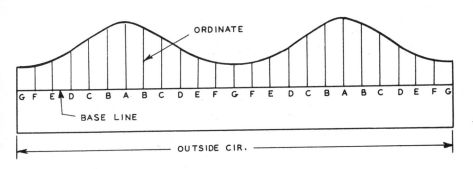

TEMPLATE FOR BRANCH

Fig. 31

the branch pipe. These lines from the base line to the end of the branch pipe will establish the lengths of the ordinate lines for the branch template.

How to Lay Out the Branch Template

Lay out a base line equal to the outside circumference of the branch pipe as shown in Figure 31. Divide this line into twenty-four equal parts (four times the number of parts on the quarter circle). Letter each point on the base line as shown. Project ordinate lines vertically from each point at convenient lengths.

Measure the length of ordinate A from the base line to the end of the branch pipe in the end view in Figure 30, and lay out this length on the ordinate A lines on the branch template (Fig. 31). Repeat this process on the ordinate lines B, C, D, E, F, and G. Connect these points with a curved line to complete the template.

How to Lay Out the Opening Template

Figure 32 will not be drawn as a separate drawing, but it will be superimposed (laid over) on Figure 30. The author has shown it here as two separate drawings for clarity.

On the end view of Figure 32, divide the quarter circle on the outside of the header from points 1 to 7 into six equal parts. Number each point as shown. Project points 1, 2, 3, 4, and 5 to the left to the front view so they cross the opening as shown. The length of these lines across the opening will establish the lengths of the ordinates for the opening template (Fig. 33). Draw a base line on the branch center line.

Lay out a straight line for the base line for the template (Fig. 33). Starting at ordinate line 1 in the center, lay out the spacings for ordinates 2, 3, 4, and 5 each way. These spacings are equal to $\frac{1}{24}$ of the outside circumference of the header. To locate point P on each end of the template, take a pair of dividers and set them for the space between 5 and P from the end view in Figure 32. Lay out this measurement from ordinate line 5 on each end of the template.

The point where line 1 crosses the opening on the front view (Fig. 32) will establish the length of ordinate 1 on the template for the opening; with your dividers, take half of the length of ordinate 1, from the base line to the edge of the opening, and lay out this length

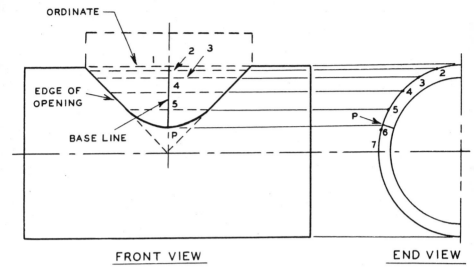

Fig. 32

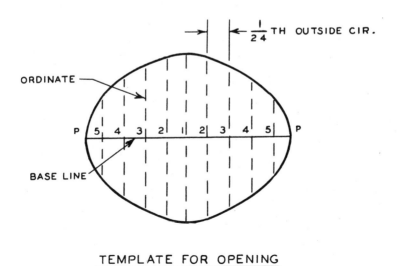

TEMPLATE FOR OPENING

Fig. 33

on each side of the base line on ordinate 1 on the template (Fig. 33). Repeat this for the lengths of ordinates 2, 3, 4, and 5. Connect the ends of the ordinate lines and points P with a curved line to complete the templates for the opening.

When cutting with the cutting torch, make a radial cut on the end of the branch pipe and the opening and bevel the edges of the opening and part of the branch pipe wherever necessary.

REDUCING TEE — BRANCH ENTERS HEADER

Lay out a circle for the outside diameter of the header, and another for the inside diameter. On the center line, above these circles, scribe a semicircle with a diameter equal to the outside diameter of the branch pipe as shown in Figure 34. Divide this semicircle into eight equal parts. Letter these points from A to E as shown. Project each point down to the inside wall of the header. This will establish the end of the branch pipe. Lay out a base line anywhere between the semicircle and the header pipe.

How to Lay Out the Branch Template

Lay out a base line with a length equal to the outside circumference of the branch pipe, as shown in Figure 35. Divide this base line into twice the number of divisions of the semicircle on the branch pipe. In this case sixteen divisions are used. Letter each of these points on the base line as shown, and project a vertical ordinate line, at a convenient length, from each point.

Measure the length of ordinate A from the base line to the end of the pipe in Figure 34, and transfer this measurement to ordinate lines A on the branch template and mark the lengths with a dot. Repeat this with ordinates B, C, D, and E. Connect these points with a smooth curved line, using an irregular curve, to complete the template. Allow a margin of an inch or more below the base line to aid in lining up the template on the pipe.

How to Lay Out the Template for the Opening

Figures 36 and 37 will be combined into one drawing, but for clarity the author has drawn them separately.

In Figure 36 lay out the front view by projecting the outside diameter of the header from the end view of Figure 34. Lay out a semicircle on the vertical center line of the branch. Divide this semicircle into eight equal parts. Letter each division from A to E as shown on the front view. Project each division down at convenient lengths. Points E will be projected down to the outside of the header establishing the side of the branch pipe.

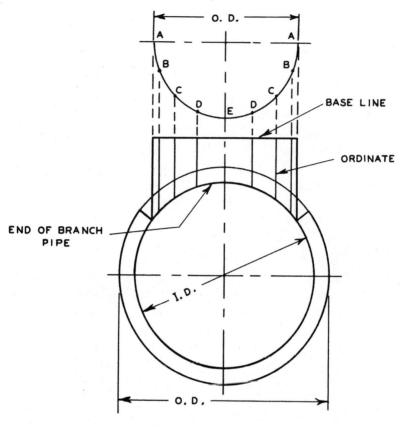

Fig. 34

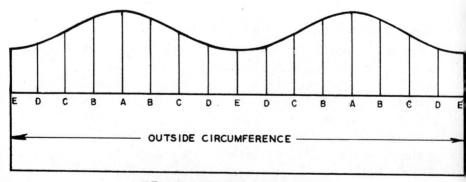

TEMPLATE FOR BRANCH

Fig. 35

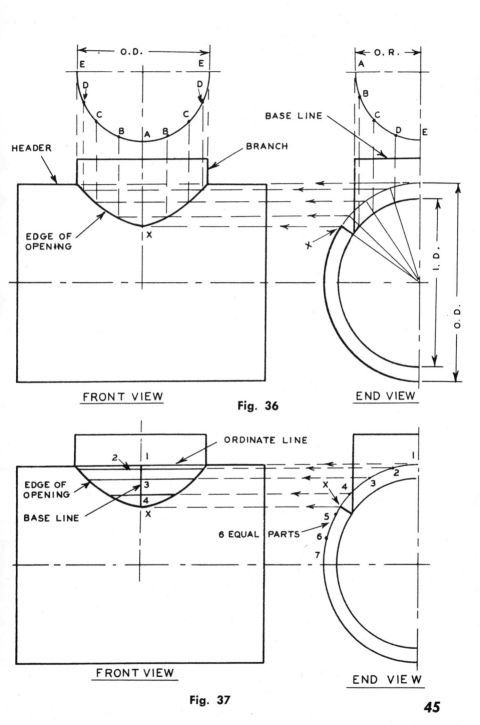

Fig. 36

Fig. 37

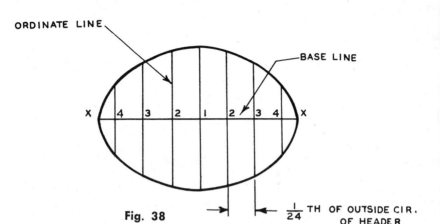

Fig. 38

TEMPLATE FOR OPENING

As shown in the end view of Figure 36, where line D from the semicircle intersects the inside wall of the header, line this point up with a straightedge with the center of the header. Project this point to the outside of the header. Then project this point over to ordinate lines D on the front view. Where these lines cross, mark the intersections with a dot.

Repeat this on lines C, B, and A. Connect these points on the front view with a curved line, establishing the edge of the opening. Where point A is projected out onto the outside of the header on the end view, mark this point with an X.

On Figure 37 on the end view, divide the quarter circle of the outside of the header into six equal parts, numbering these points from 1 to 7 as shown. Each curved space between numbers is equivalent to one twenty-fourth of the outside circumference of the header. Project points 1, 2, 3, and 4 across the opening on the front view as shown.

Figure 38 shows the template for the opening. Lay out a base line as shown. Starting at the center at line 1, lay out the spacings, 1 to 2, 2 to 3, and 3 to 4, to the right and left. To obtain the exact length of these spacings, divide the outside circumference of the header by twenty-four. On each end lay out the distance from 4 to X. This measurement is taken off the end view of Figure 37 with a pair of dividers from 4 to X.

Use the branch center line on the front view in Figure 37 as the base line for the opening. Where line 1 crosses the opening, set the dividers from the base line to the edge of the opening, and lay out this measurement on line 1 on each side of the base line on the template (Fig. 38). Repeat this with lines 2, 3, and 4. Connect the ends of these ordinate lines and points X with a smooth curved line to complete the template.

When cutting, make a radial cut on the branch and opening; then bevel the opening. A radial cut is made by pointing the cutting torch tip to the center of the pipe at all times. Do not bevel the branch.

REDUCING TEE
WITH BRANCH PIPE OUTSIDE HEADER

In Figure 39 lay out a circle for the outside diameter of the header, and draw a vertical line through the center. On the center line above the circle, scribe a semicircle with a diameter equal to the outside diameter of the branch pipe. Divide the semicircle into eight equal parts. Letter each point from A to E as shown. Project each point down to the outside wall of the header. This will establish the end of the branch pipe. Lay out a base line anywhere between the semicircle and the header pipe as shown.

How to Lay Out the Branch Template

Lay out a base line with a length equal to the outside circumference of the branch pipe, as shown in Figure 40. Divide this base line into twice the number of divisions of the semicircle of the branch pipe. In this case the base line is divided into sixteen equal parts since there are eight parts on the semicircle. Letter each of these points as shown, and project a vertical ordinate line from each, at a convenient length.

Measure the length of ordinate A from the base line to the end of the pipe in Figure 39, and transfer this measurement to the ordinates A on the branch template and mark the lengths with a dot. Repeat this with ordinates B, C, D, and E. Connect these points with a curved line, using an irregular curve. To complete the template, allow a margin of an inch or more below the base line to aid in lining up the template on the pipe.

How to Develop the Opening Template

In Figure 41 lay out the end view by first laying out a vertical and a horizontal center line. Then lay out a large semicircle with a diameter equal to the outside diameter of the header pipe. Lay out another circle equal to the inside diameter of the header. On the center line above the header, scribe a quarter circle with a radius equal to the inside radius of the branch pipe. Divide this quarter circle into four equal parts. Number each point from 1 to

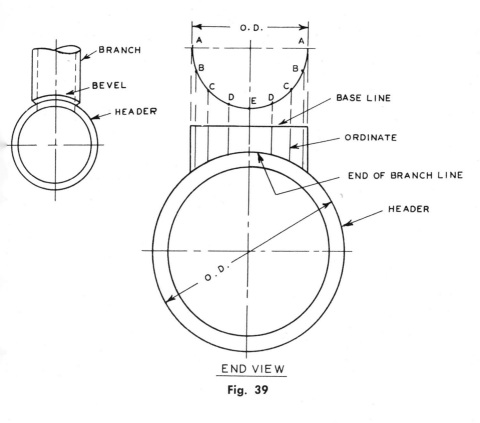

END VIEW
Fig. 39

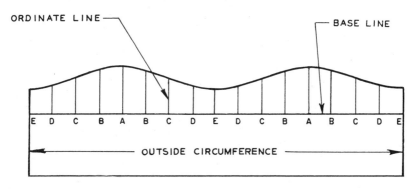

BRANCH TEMPLATE
Fig. 40

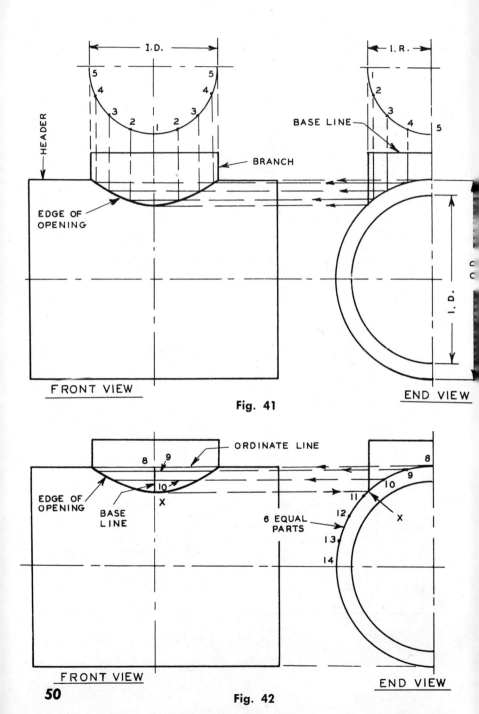

Fig. 41

Fig. 42

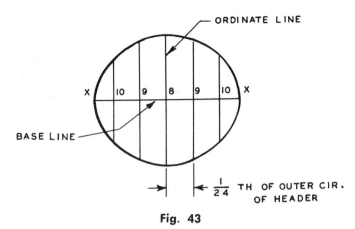

Fig. 43

TEMPLATE FOR OPENING

5 as shown. Project each division down to the outside of the header pipe.

Develop the front view by projecting the outside of the pipe from the end view for the header. Erect a vertical center line, and scribe a semicircle with a diameter equal to the inside diameter of the pipe. Divide the semicircle into eight equal parts. Number each division from 1 to 5 as shown, and project both of the number 5 points down to the top of the header. Project points 1 to 4 down at convenient lengths.

To develop the edge of the opening curve line, where the number 4 line meets the outside of the header pipe on the end view, project to the left to the intersection of the number 4 lines on the front view, and mark the intersections with a dot. Repeat this with points 3, 2, and 1. Connect the dots for the edge of the opening on the front view, using an irregular curve.

Figure 42 will actually be drawn on Figure 41 but for clarity the author has made it a separate drawing. Divide the quarter circle from point 8 to point 14 into six equal parts on the end view, numbering each point from 8 to 14 as shown. Project point 9 to the left until it crosses the curved "edge of opening" line. Project point 10 across in the same manner. Point 11 cannot be projected across since it will be below the "edge of opening" line, so mark the low point of the edge of the opening on the front view as X. Project this point to the right to the semicircle; this will be located between points 10 and 11. Mark this point as X. NOTE: this point will vary as the branch pipe increases in size. It may fall between points 11 and 12 or 12 and 13 on other developments. Keep this in mind: the location of the X point will vary as the branch pipe increases in size.

The curved spacings from 8 to 9, 9 to 10, and 10 to X on the end view will be the spacings of the ordinates on the opening template in Figure 43. The spacings between 8 and 9 and 9 and 10 can be taken off the end view by bending a piece of paper to the curvature of the header circle and marking them; or each space may be made to equal $\frac{1}{24}$ of the outer circumference. Divide the circumference by twenty-four for the accurate spacing length.

On Figure 43 lay out a base line and in the center draw a vertical line for ordinate 8. Lay out the spacings for ordinates 9 and 10 as explained above. With a pair of dividers, take the distance from point 10 to point X from the end view in Figure 42, and lay out this measurement on the base line from ordinate line 10 on each end of the opening template.

On the front view of Figure 42, the points at which ordinates 8, 9, and 10 cross the edge of the opening curved line will give the lengths of the ordinates in Figure 43. To transfer these lengths, first draw a base line on the center line as shown. Then with a pair of dividers or a compass take half of the length of ordinate 8 (distance from the base line to the edge of the opening), and lay this length out on each side of the base line for the length of ordinate 8 in Figure 43. Repeat this with ordinates 9 and 10. Connect the ends of the ordinate lines and points X with a curved line to complete the opening template.

Radial-cut the branch pipe with the cutting torch; then bevel the edge.

When cutting the opening, keep the cutting tip parallel with the center line at all times. Do not bevel the opening edge.

ECCENTRIC TEE

Lay out a large circle with a diameter equal to the inside diameter of the header. To the right lay out a semicircle with a diameter equal to the outside diameter of the branch pipe. Keep the bottom of the two circles on the same elevation. See Figure 44. Divide the smaller semicircle into eight equal parts. Number each point from 1 to 9 as shown. Extend each point to the larger circle. Lay out a base line anywhere between the header circle and the branch semicircle.

How to Lay Out the Branch Template

Lay out a base line with a length equal to the outside circumference of the branch pipe. Divide the base line into 16 equal parts. Number the division points from 1 to 9 as shown in Figure 45. Extend ordinate lines up from each of the division points.

From Figure 44 transfer the length of ordinate 9 (distance from the base line to the larger circle) to ordinate 9 line on the branch template, and mark the point with a dot. See Figure 45. Repeat this with ordinates 1, 2 ,3, 4, 5, 6, 7, and 8. Connect these points on the branch template with a curved line, using an irregular curve (French curve) to do so. Allow a margin of one inch or more below the base line for line-up purposes when placing the template on the pipe for marking the cut line.

How to Lay Out the Opening Template

For clarity the author has separated Figure 46 from Figure 44 to show how the ordinate lengths are obtained. To save time in drawing, Figure 46 would be superimposed upon Figure 44.

Figure 46 shows a large circle with a diameter equal to the inside diameter of the header pipe. Divide the right half of the circle into eight equal parts. Letter each point from *A* to *I*.

To the right lay out a semicircle with a diameter equal to the outside diameter of the branch pipe. Project the top of the semicircle to the larger circle, and mark this point with an *X*. Figure 46 shows *X* located between points *F* and *G*. This location will vary as the branch pipe size increases or decreases in diameter.

Project points *A*, *B*, *C*, *D*, *E*, and *F* to the center line of the semicircle. The length of the lines from the circumference of the semi-

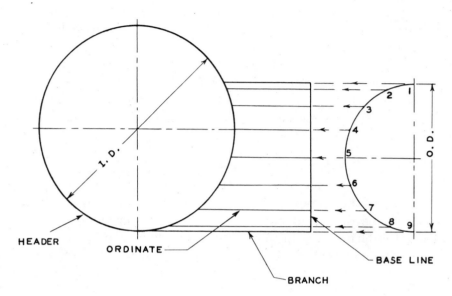

Fig. 44

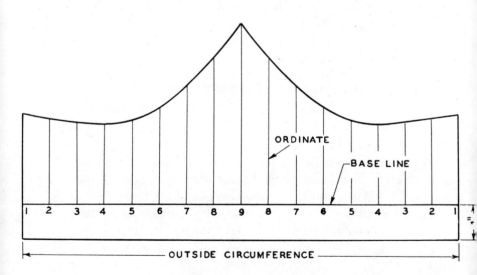

BRANCH TEMPLATE

Fig. 45

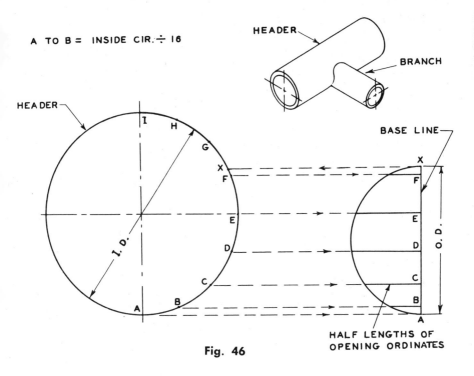

Fig. 46

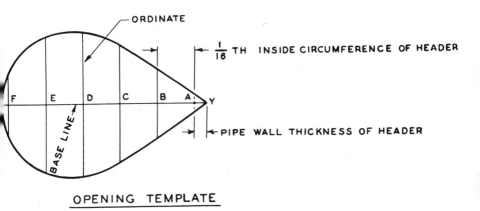

OPENING TEMPLATE

Fig. 47

circle to the center line will be half the length of each ordinate for the opening template. The center line will be the base line.

Lay out a base line as shown in Figure 47. The spacings of the ordinate lines are equal to the spacings between the lettered points on the header circle of Figure 46. To determine the accurate length of the spacings, divide the inside circumference of the header by twice the number of divisions on the header half circle. In this case, divide by sixteen since there are eight spacings on the half circle.

Lay out these spacings on the base line for ordinate lines A to F, and draw random length lines at right angles for ordinate lines B to F.

From the header circle in Figure 46, with a pair of dividers, take the distance from F to X, and lay out this length on the base line from F, locating X in Figure 47.

On the base line from point A, lay out a length equal to the thickness of the header pipe wall, locating point Y.

With a pair of dividers, take the length of ordinate B from the branch semicircle in Figure 46, and lay out this length on each side of the base line on Figure 47. Repeat this with ordinates C, D, E, and F.

Ordinate A has no given length. Where the ends of ordinate B connect with point Y will establish the length of ordinate A. Connect the X and Y points with the ends of the ordinate lines with a curved line, using the irregular curve (French curve), to complete the template.

This explanation uses sixteen divisions on the branch and the header. On larger pipes, use more divisions for accuracy. Bevel the opening only.

BRANCH PIPE FROM THE BACK OF A WELDING ELBOW

Whether the branch pipe is eccentric (off center) or concentric (on center), the layout is the same. The end of the branch pipe will fit against the outside of the elbow.

In Figure 48, using point K as a compass point, scribe arcs, laying out the outside walls of the elbow as shown.

Lay out a branch center line and at any convenient distance from the elbow, locate point L. Using this point, scribe a semicircle with a diameter equal to the outside diameter of the branch pipe. Divide this semicircle into six equal parts. Number each point from 1 to 7 as shown and extend each division upwards at a convenient length. Draw a base line anywhere between the elbow and the semicircle.

To locate compass point M, draw a straight line from the face of the elbow and extend the center line running through compass point L until the two lines intersect. From point M scribe a quarter circle with a radius of half the outside diameter of the pipe. Divide the quarter circle into four equal parts. Number each point as shown, 1–7, 2–6, 3–5, and 4. The numbers are projected over from the semicircle.

Scribe a semicircle from point P on the center line of the elbow equal to the outside diameter of the elbow. Project points 1–7 up to this semicircle from the quarter circle. Project points 2–6 up to the semicircle; project this point of intersection to the face line. Set the compass on point K, and open it up to the 2–6 point on the face line. Scribe an arc until it intersects ordinate lines 2 and 6 on the branch pipe. This procedure will determine the lengths of ordinates 2 and 6 for the branch template.

From the quarter circle project points 3–5 up to the semicircle, and project to the left to the face line. Set the compass at point K, open it up to points 3–5 on the face line, and swing an arc until it intersects ordinates 3 and 5 on the branch. Repeat this process with point 4 on the quarter circle.

Connect the intersection of the arcs and the straight ordinate lines of the branch with a curved line, to establish the end of the branch pipe.

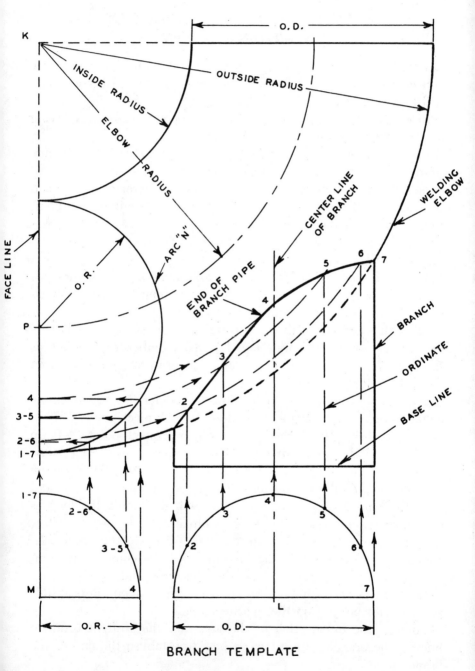

BRANCH TEMPLATE

Fig. 48

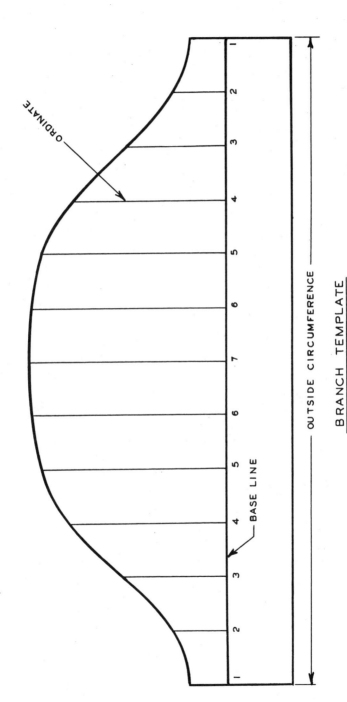

BRANCH TEMPLATE

Fig. 49

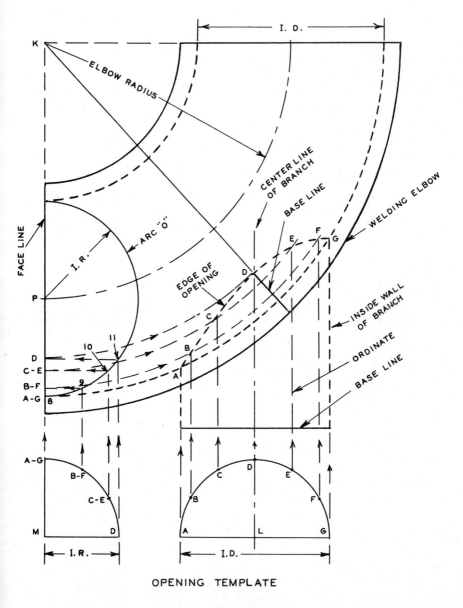

OPENING TEMPLATE

Fig. 50

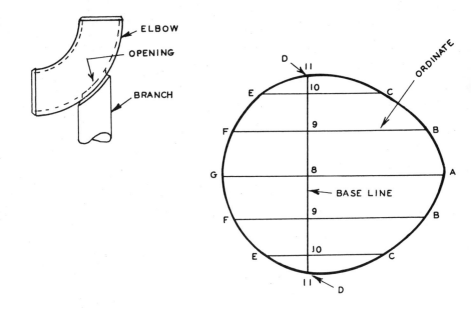

TEMPLATE FOR OPENING

Fig. 51

How to Lay Out the Branch Template

Lay out a base line with a length equal to the outside circumference of the branch pipe as shown in Figure 49. Divide this base line into twelve equal parts, twice the number of parts as shown on the branch semicircle in Figure 48. Number each point from 1 to 7 as shown. Extend ordinate lines upward from each point at a convenient length.

Ordinate lengths are obtained from Figure 48 with a pair of dividers or a compass. On ordinate line 1 take the distance from the branch base line to the curve representing the end of the branch pipe, and transfer this length to the number 1 ordinate lines on the branch template (Fig. 49), laying the length out from the base line and marking each length with a dot.

Take the length of ordinate line 2 from the base line to the end of the branch curve from Figure 48, and lay out this length on ordinate lines 2 on the branch template (Fig. 49), marking each length with a dot. Repeat this with the 3, 4, 5, 6, and 7 ordinate lengths. Connect each dot with a curved line to complete the template.

How to Lay Out the Opening Template

For clarity the author has separated Figures 48 and 50 to show how the ordinate lengths for the opening template are obtained. To save time, Figure 50 would be superimposed over Figure 48.

In Figure 50, using point K, scribe arcs laying out the outside and inside walls of the elbow as shown. Lay out a branch center line in the same position as in Figure 48. Also locate compass point L in the same position.

Using compass point L, scribe a semicircle with a diameter equal to the inside diameter of the branch pipe, and divide this semicircle into six equal parts. Letter each point from A to G as shown. Extend each division upwards at a convenient length.

With compass point M located in the same position as in Figure 48, scribe a quarter circle with a radius of half the inside diameter of the branch pipe. Divide the quarter circle into four equal parts. Letter each point as shown, A–G, B–F, C–E, and D. Letters are projected over from the semicircle.

Scribe a semicircle from point P on the center line of the elbow, equal to the inside diameter of the elbow. Project points A–G up to this semicircle from the quarter circle. Project points B–F up to the semicircle; and, where it intersects, project to the left to the face line. Set the compass on point K, open it up to the B–F point on the face line, and scribe an arc until it intersects ordinate lines B and F on the branch pipe.

From the quarter circle, project points C–E up to the semicircle, and project to the left to the face line. Set the compass at point K, and open it up to points C–E on the face line. Swing an arc until it intersects ordinate lines C and E on the branch pipe. Repeat this with point D on the quarter circle.

Connect the intersections of the arcs and straight ordinate lines with a dotted curved line, to establish the edges of the opening. Letter each intersection as shown.

Line up point K and point D on the opening with a straightedge, and draw the base line as shown.

Figure 51 shows the opening template. Lay out a base line at a convenient length, and draw the line G–A at a convenient length. The point at which line G–A intersects the base line is numbered 8. On the semicircle in Figure 50 take the curved length from point 8 to point 9. (This can be done by bending a piece of paper to the same curvature and marking it.) Lay out this length on the base line of Figure 51 from point 8, locating point 9.

Draw ordinate lines through points 9 at convenient lengths. This procedure will establish the F–B lines. In Figure 50 measure the curved length from point 9 to point 10 on the semicircle, and lay out this length on the base line of Figure 51 from points 9, locating points 10. Draw ordinate lines through points 10 at convenient lengths. These will be the C–E lines.

In Figure 50 measure the curved length from point 10 to point 11 on the semicircle, and lay out this length on the base line of Figure 51 from points 10, locating points 11 or D.

The ordinates for the opening template (Fig. 51) are the curved line lengths A to G, B to F, and C to E from Figure 50. To obtain these lengths, take a stiff piece of paper, and mark a point representing the base line. Bend the paper to the same curvatures as the A–G line; then line up the base line point on the paper with the base line on the drawing, and mark the points A and G on the paper.

Lay this paper on the A–G line of Figure 51. Line up the base line on the paper with the base line of the template, and mark points A and G on the line with a dot. With the stiff paper, obtain the lengths of the B–F and the C–E lines from Figure 50 in the same manner as curved line A–G was obtained, and transfer the lengths to Figure 51, locating points B and F and C and E. Points D will have no length, so they will be located at points 11 on each end of the base line. Connect all the lettered points with a curved line to complete the opening template.

When cutting the branch pipe, scarf the end of it to fit the curvature of the outside surface of the elbow; then bevel the edge.

When cutting the hole, keep the cutting tip pointed straight in or parallel with the branch center line. Do not bevel the edge of the hole.

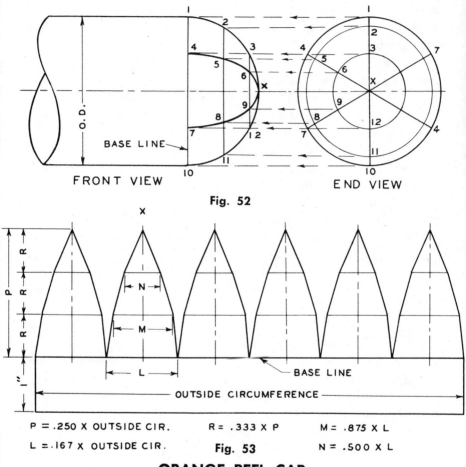

Fig. 52

P = .250 X OUTSIDE CIR.   R = .333 X P   M = .875 X L
L = .167 X OUTSIDE CIR.   Fig. 53   N = .500 X L

ORANGE PEEL CAP

To make a template for an orange peel cap, lay out a circle with a diameter equal to the outside diameter of the pipe as shown on the end view in Figure 52. Divide this circle into six equal parts. Number these points 1, 4, 7, and 10 as shown.

On the front view lay out the base line 1–10 at any convenient point and scribe a half circle, using the intersection of the center line and the base line as the compass point. This semicircle represents the end of the pipe. Project the sides of the pipe to the left of the semicircle. Divide the quarter circle from 10 to X into three equal parts, locating points 11 and 12. At point 11 draw a light line across the semicircle. At point 12 draw a light line across the semicircle. Set the compass from the center line to point 11, on the front view; and with this radius, scribe a light circle on the end view. Set the compass from the center line to point 12 on the front view and with this radius scribe a light circle on the end view. Number the points where the circles cross the straight lines as shown.

From the points where lines 4 and 7 meet the outer circle on the end view, project lines to the base line 1–10 on the front view. From the points where lines 4 and 7 cross the largest inner circle at points 5 and 8 on the end view, project lines to line 2–11 on the front view. Where lines 4 and 7 cross the smaller inner circle at points 6 and 9 on the end view, project these points to line 3–12 on the front view.

Connect points 4, 5, 6, and X together with a curved line, using an irregular curve, on the front view. Repeat this with points 7, 8, 9, and X, establishing the outside appearance of the orange peel cap.

How to Lay Out the Template

Lay out the base equal to the outside circumference of the pipe as shown in Figure 53. Divide this line into six equal parts for six arms. Dimension L equals one sixth of the outside circumference. Lay out dimension P the length of the arm, which is equal to one fourth of the outside circumference of the pipe, locating point X. Divide the length of each arm into three equal parts as shown, and draw lines at right angles to the center of each arm. Dimension M equals ⅞ of dimension L. Lay out half of M on each side of the center line as shown. N equals ½ of dimension L. Lay out half of N on each side of the center line as shown. Connect the ends of lines L, M, N, and point X with straight lines to complete one arm. A curved line could also be used in place of the straight lines. Repeat this on the other five arms.

Allow a margin of 1 in. or more below the base line for line-up purposes when the template is placed around the pipe.

Make a radial cut when using the cutting torch; then bevel the edge of each arm before folding the ends together.

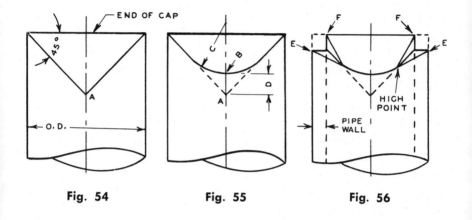

Fig. 54   Fig. 55   Fig. 56

BULL PLUG CAP

Figures 54, 55, and 56 show each step taken to develop the lines. These will not be drawn separately. All three will be combined into one drawing as shown in Figure 57.

To begin, lay out a center line as shown in Figure 54, and lay out half the outside diameter of the pipe on each side. Then lay out the line for the end of the pipe wherever required. From the corners, draw the lines to the center on a 45-deg. angle, locating point A. From point A on the center line, measure distance D, which is equal to 2 times the pipe-wall thickness, locating point B as shown in Figure 55.

Join the 45-deg. lines and point B with a smooth curved line. This is done with a compass. By trial and error, locate a compass point on the center line. The setting on the compass will be radius C.

Lay out the dotted lines for the inside wall of the pipe as shown in Figure 56. Where the inside walls intersect the 45-deg. lines, draw the short lines horizontally and vertically, locating points E and F.

Line up point E with a straightedge to the high point of the curve line and draw a straight line. Connect the intersection of this line and the high point of the curve with point F. Repeat this process on the opposite side.

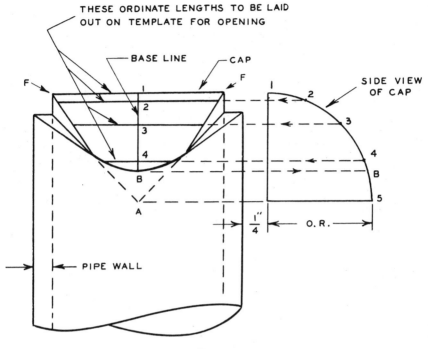

Fig. 57

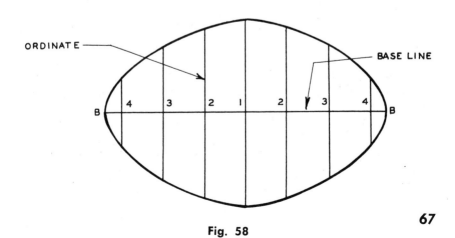

Fig. 58

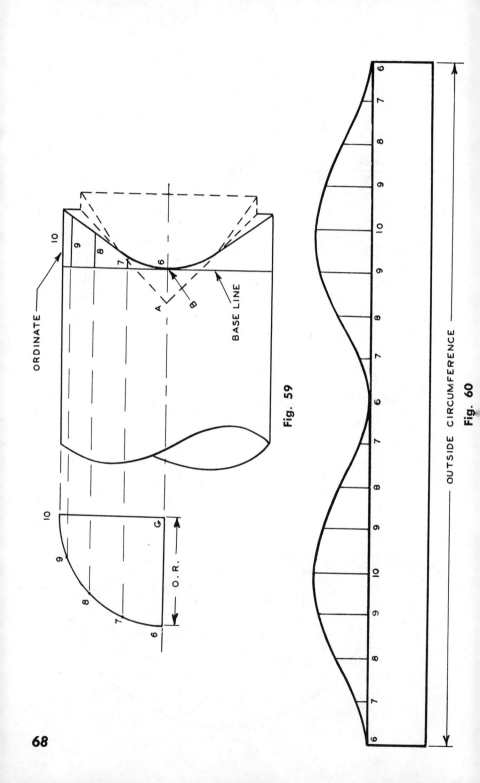

Fig. 59

Fig. 60

How to Lay Out the Cap Template

As shown in Figure 57, extend point A to the outside of the pipe, and about a quarter inch from the side of the pipe, scribe a quarter circle with a radius equal to the outside radius of the pipe. Divide this quarter circle into four equal parts, numbering the points from 1 to 5 as shown. Each spacing on the curve between numbers is equal to $1/16$ of the outside circumference. Extend point B over to the quarter circle as shown between 4 and 5. Project points 1, 2, 3, and 4 across the cap as shown. These lines, where they cross the edges of the cap, will be the ordinate lines for the cap template. Draw a solid line on the center line as the base line.

Lay out a base line as shown in Figure 58, and lay out the spacings from 1 to 2, 2 to 3, and 3 to 4. These spacings are equal to $1/16$ of the outside circumference, or the curved line spacing between the numbers on the quarter circle of Figure 57. With a pair of dividers take the spacing from 4 to B from the quarter circle in Figure 57, and lay out this space from ordinates number 4 on the base line in Figure 58, locating points B. Draw the ordinate lines at right angles to the base line at convenient lengths.

With a pair of dividers or a compass, measure one half of the length of ordinate 1 (from the base line to corner F) on the cap in Figure 57, and lay this measurement out on ordinate line 1 on each side of the base line on the template as shown in Figure 58. Measure ½ of the length of ordinate 2 (from the base line to the edge of the cap) in Figure 57, and lay this measurement out on ordinate lines number 2 on each side of the base line on the template. Repeat this with ordinate lines 3 and 4. Connect the ends of the ordinate lines and points B with a smooth curved line (using an irregular curve) to complete the template.

How to Lay Out the Template For the End of the Pipe

Figure 59 will not be a separate drawing, but will be superimposed on Figure 57. The author has shown it here as a separate drawing for clarity.

At any convenient point on the center line, locate a compass point G, and scribe a quarter circle with a radius equal to ½ the outside diameter of the pipe. Divide this quarter circle into four equal parts. Number these points from 6 to 10 as shown. Through point B, draw a base line. Extend point 7 from the quarter circle, and draw a solid line from the base to the first slope line as shown. Repeat this with points 8 and 9. These lines from the base to the first slope line will be the ordinate lengths for the template shown in Figure 60.

In Figure 60 lay out a base line equal to the length of the outside circumference of the pipe. Divide this base line into sixteen equal parts. Number these points as shown. Extend ordinate lines from these points at right angles to the base at convenient lengths.

Measure the length of ordinate 7 in Figure 59, and lay this measurement out on ordinate lines number 7 on the template as shown in Figure 60. Repeat this with ordinate lengths 8, 9, and 10. Ordinate number 6 will have no length since it is on the base line in Figure 59. Therefore, it will be on the base line of the template. Connect the ends of the ordinate lines with a smooth curved line, using an irregular curve. Drop down about an inch or more from the base line to provide a margin at the bottom of the template for lining-up purposes when wrapping the template around the pipe.

Make a radial cut with the cutting torch for the cap and the end of the pipe, and bevel wherever necessary.

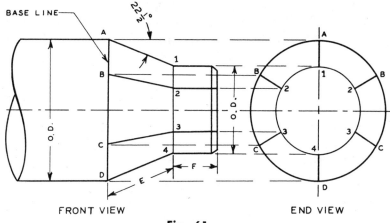

Fig. 61

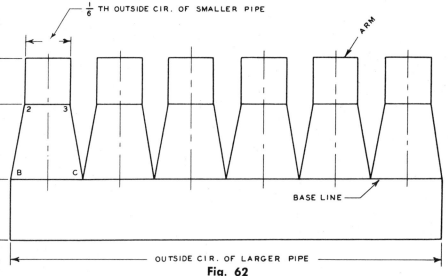

Fig. 62

CONCENTRIC REDUCER

On the end view in Figure 61, lay out a circle for the outside diameter of the larger pipe, and a circle for the outside diameter of the smaller pipe. Divide each circle into six equal parts. Letter and number these points A, B, C, and D, and 1, 2, 3, and 4 as shown. Connect these points with straight lines as shown.

71

Project these two circles to the front view. Draw the base line A–D at any convenient point. At points A and D draw lines on a 22½-deg. angle to the outside of the smaller pipe, locating points 1 and 4. Draw a line connecting points 1 and 4. Dimension F equals 1 in. Lay out dimension F from line 1–4, establishing the end of the pipe.

Project points B and C from the side view to the base line A–D on the front view. Project points 2 and 3 from the end view to line 1–4 on the front view. On the front view connect points B and 2 with a straight line. Extend point 2 to the end of the pipe. Connect points C and 3 with a straight line, and extend point 3 to the end of the pipe. These lines represent the meeting points of the arms when they are folded together. The length of the slope from D to 4 is indicated as dimension E and represents the true lengths of the arms.

How to Lay Out the Template

Lay out a base line equal to the outside circumference of the larger pipe, as shown in Figure 62. Divide this base line into six equal parts. The distance from B to C equals ⅙ of the outside circumference of the larger pipe. Lay out a center line for each arm.

From the base line, on the center line, lay out dimension E, which is obtained from Figure 61, and draw a line parallel with the base line. On this line lay out 1/12 of the outside circumference of the smaller pipe on each side of the center line, locating points 2 and 3. Connect points 2 and B and points 3 and C with a straight line.

From line 2–3 lay out dimension F, locating the end of the arm. Draw a line parallel with line 2–3. Extend points 2 and 3 to this line, completing one arm. Repeat this on the other five arms since they are all alike. Provide a margin of 1 in. below the base line for line-up purposes when the template is wrapped around the pipe. Radial-cut each arm and bevel the edges before folding them together.

ECCENTRIC REDUCER

On the end view of Figure 63 lay out a semicircle with a radius equal to the outside radius of the larger pipe, and divide the semicircle into eight equal parts. Letter these points from A to K as shown. Then lay out a semicircle with a radius equal to the outside radius of the smaller pipe, with the two circles touching at the bottom at point K. Divide this semicircle into eight equal parts, and number these points from 1 to 9 as shown.

Connect points B and 2, D and 4, and F and 6 with a straight solid line. These lines represent the meeting points when the arms are folded together. Points A, C, and E are the centers of the arms at the larger end.

Connect points B and 3, D and 3, D and 5, and F and 5 with dotted lines. Points 1, 3, and 5 are the centers of the arms at the smaller end.

Project the outside diameters of the two pipes to the front view. Lay out the base line A–K at any convenient point where you want the slope line to start. From point A project a 30-deg. line down until it intersects the outside of the smaller pipe at point 1. Draw line 1–9. Lay out a dimension of one inch from this line locating the end of the pipe. If a longer taper is desired, use a 22½-deg. angle.

Project points B, D, and F from the end view to the base line A–K on the front view. Project points 2, 3, 4, 5, and 6 from the end view to the 1–9 line on the front view. On the front view connect points B and 2, D and 4, and F and 6 with a solid straight line. Project the solid lines from points 2, 4, and 6 to the end of the pipe. Connect points B and 3, D and 3, D and 5, and F and 5 with dotted lines.

The solid and dotted lines on the front view are not true lengths due to the sloping in of the pipe from the larger to the smaller diameter. To lay out the true lengths of these lines, triangles must be formed, and the hypotenuse of each triangle will be the true length of the lines between the numbers and letters.

To lay out the true lengths of these lines, measure the length of the line from K to 9, on the front view in Figure 63, and lay this length out vertically as shown in Figure 64. Draw a base line at point 9.

To lay out the true length of the solid line F–6, measure the

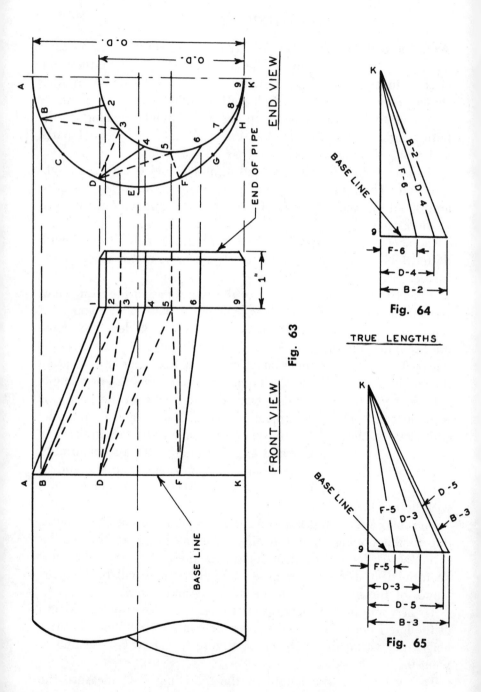

length of the solid line *F* to 6 from the end view in Figure 63, and lay out this length from point 9 on the base line in Figure 64. Connect this point on the base line with *K* on the vertical line. This is the hypotenuse of a right angle triangle, and is the true length of line *F*–6. On the base line, lay out the length *D*–4 and *B*–2. These lengths are obtained from the end view. Connect these points on the base line with point *K* for the true lengths of lines *D*–4 and *B*–2.

To lay out the true lengths of the dotted lines, lay out the length of the line from *K* to 9 vertically and draw a base line at point 9 as shown in Figure 65.

Obtain the length of the dotted line from *F* to 5 from the end view in Figure 63, and lay out this length on the base line from point 9 in Figure 65. Connect this point on the base line with *K* on the vertical line. The hypotenuse of this right-angle triangle is the true length of line *F*-5. Transfer the dotted lengths *D*–3, *D*–5, and *B*–3 from the end view to the base line in Figure 65, and connect these points with *K* for the true lengths of lines *D*–3, *D*–5, and *B*–3.

How to Lay Out the Template

$S = 1/8$ of the outside circumference of the smaller pipe.

$X = 3/16$ of the outside circumference of the smaller pipe.

Lay out a base line equal to the length of the outside circumference of the larger pipe. Divide this base line into sixteen equal parts. Letter these points as shown in Figure 66. At point *A* erect a vertical line. On this line lay out the length from *A* to 1. This length is obtained from points *A* to 1 on the front view of Figure 63. Using point 1 as a compass point, set the compass for radius *P* which is one half of *S*, and scribe the two small arcs as shown. Set the compass for the true length of line *B*–2 (from point *K* to the base line) from Figure 64. With this compass setting, set the compass on points *B* on the base line and scribe arcs so they cross the smaller arcs. Connect the intersection of these arcs with a straight horizontal line. Then connect the intersections of these arcs with straight lines to points *B* on the base line. Project lines vertically from the intersection of the arcs for a length of 1 in. Draw a horizontal line connecting these two lines. This line represents the end of the arm.

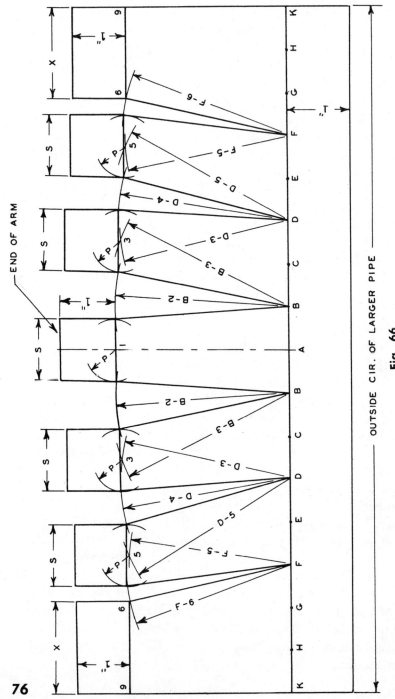

Since both arms with the centers shown as 3 are laid out the same, only one will be explained. Set the compass for the length of line B–3 in Figure 65, and with this setting set the compass on B on the base line in Figure 66 and scribe an arc. With the compass setting equal to the length of line D–3 in Figure 65, set the compass on point D on the base line in Figure 66 and scribe an arc. The intersection of these two arcs will locate point 3, the center of the arm. With point 3 as the compass point, scribe the two small arcs with a radius of P. Set the compass for the radius D–4 in Figure 64; then set the compass on point D on the base line (Fig. 66) and scribe an arc so it intersects the small arc.

Connect the intersections of these arcs with a straight horizontal line; then connect the intersections with B and D on the base line with straight lines. Project lines vertically from the arc intersections for a length of 1 in.; then draw a horizontal line connecting these two lines for the end of the arm.

Both arms with the centers shown as 5 are laid out the same so only one will be explained. From Figure 65 set the compass for the length of line D–5, and with this setting, set the compass on point D on the base line and scribe an arc. With the compass setting equal to the length of line F–5 from Figure 65, set the compass on F on the base line and scribe an arc. Where these two arcs intersect will locate point 5, the center of the arm. With point 5 as a compass point, scribe the two small arcs with a radius of P.

Set the compass for the radius F–6 (Fig. 64); then set the compass on point F on the base line (Fig. 65) and scribe an arc intersecting the small arc. Connect the intersections of the arcs with a straight horizontal line; then connect the intersections with D and F on the base line with straight lines. Project lines vertically from the arc intersections for a length of 1 in.; then draw a horizontal line connecting these two lines for the end of the arm.

Both of the end arms are laid out the same so only one will be explained. To lay out the two end arms, take the length of the line from K to 9 from the front view of Figure 63, and lay out this length vertically from point K on the base line, locating point 9. From point 9 lay out a vertical line 1 in. long; then draw a horizontal line to establish the end of the arm. Draw a horizontal line from point 9 and lay out dimension X, locating point 6. Draw a vertical line from point 6 to the end of the arm. If your template is accurate, the arc drawn with radius F–6 should intersect point 6. Connect point 6 with a straight line to point F on the base line. Complete the other half of the template in the same manner.

Lay out a margin of an inch or more from the base line on the bottom of the template for line-up purposes when the template is wrapped around the pipe.

Make a radial cut on each arm when using the cutting torch. Bevel the edges of the arms before folding together.

DECIMAL EQUIVALENTS OF FRACTIONS OF AN INCH

FRACTION OF AN INCH			DECIMAL OF AN INCH	FRACTION OF AN INCH			DECIMAL OF AN INCH
1/64			.015625	33/64			.515625
	1/32		.03125		17/32		.53125
3/64			.046875	35/64			.546875
		1/16	.0625			9/16	.5625
5/64			.078125	37/64			.578125
	3/32		.09375		19/32		.59375
7/64			.109375	39/64			.609375
		1/8	.125			5/8	.625
9/64			.140625	41/64			.640625
	5/32		.15625		21/32		.65625
11/64			.171875	43/64			.671875
		3/16	.1875			11/16	.6875
13/64			.203125	45/64			.703125
	7/32		.21875		23/32		.71875
15/64			.234375	47/64			.734375
		1/4	.25			3/4	.75
17/64			.265625	49/64			.765625
	9/32		.28125		25/32		.78125
19/64			.296875	51/64			.796875
		5/16	.3125			13/16	.8125
21/64			.328125	53/64			.828125
	11/32		.34375		27/32		.84375
23/64			.359375	55/64			.859375
		3/8	.375			7/8	.875
25/64			.390625	57/64			.890625
	13/32		.40625		29/32		.90625
27/64			.421875	59/64			.921875
		7/16	.4375			15/16	.9375
29/64			.453125	61/64			.953125
	15/32		.46875		31/32		.96875
31/64			.487375	63/64			.984375
		1/2	.5			1	1.000000

STANDARD PIPE DATA

NOMINAL PIPE DIAMETER INCHES	ACTUAL INSIDE DIAMETER INCHES	ACTUAL OUTSIDE DIAMETER INCHES	ACTUAL OUTSIDE CIRCUMFERENCE INCHES
1	1.049	1.315	4.131
1¼	1.308	1.660	5.215
1½	1.610	1.900	5.969
2	2.067	2.375	7.461
2½	2.469	2.875	9.032
3	3.068	3.500	11.000
3½	3.548	4.000	12.566
4	4.026	4.500	14.137
4½	4.560	5.000	15.708
5	5.047	5.563	17.476
6	6.065	6.625	20.813
8	7.981	8.625	27.096
10	10.020	10.750	33.772
12	12.000	12.750	40.055
14	13.250	14.000	43.982
16	15.250	16.000	50.265
18	17.250	18.000	56.549
20	19.250	20.000	62.832
24	23.250	24.000	75.398
26	25.250	26.000	81.681
30	29.250	30.000	94.248
36	35.250	36.000	113.098

INDEX

Angle of cut, formula, 9
Angle plate, and pipe miter, 14

Bisecting an angle, 4
Bisecting a line, 2
Branch pipe from the back of an elbow, 57
Branch template, for angle plate and pipe miter, 14; for branch pipe from the back of a welding elbow, 61; for double angled plate and pipe miter, 19; for eccentric tee, 53; for full sized lateral, 30; for full size tee, 40; for reducing lateral, 34: for reducing tee — branch enters header, 43; for reducing tee — branch pipe outside header, 48; for true Y, 25
Bull plug, and pipe template, 69

Cap, bull plug, 66; orange peel, 64
Cap template, for a bull plug, 69
Cutting torch, 1
Cutting torch technique, for angle plate and pipe miter, 18; for branch pipe from the back of a welding elbow, 63; for bull plug cap, 70; for concentric reducer, 72; for double angled plate and pipe miter, 24; for eccentric reducer, 78; for full size lateral, 31; for full size tee, 42; for miter turns, 1, 11; for orange peel cap, 65; for reducing lateral, 37; for reducing tee — branch enters header, 47; for reducing tee — branch outside header, 52; for true Y, 27

Decimal equivalents of fractions of an inch, 79
Dividing a line, into equal parts with a rule, 4; into 6 equal parts, 3; into 4 equal parts, 3

Dividing a quarter circle, 6 equal parts, 6; 8 equal parts, 5
Divisions, number of, 11, 18, 24
Double angled plate and pipe miter, 19
Drawing equipment, 2

Eccentric tee, 53

Fractions of an inch, 79
Full size lateral, 30
Full size tee, 38

Irregular curve, 7

Miter turns, 9

Opening template, for angle plate and pipe miter, 15; for branch pipe from the back of a welding elbow, 62; for double angled plate and pipe miter, 21; for eccentric tee, 53; for full size lateral, 31; for full size tee, 40; for reducing lateral, 35; for reducing tee — branch enters header, 43; for reducing tee — branch pipe outside header, 48
Orange peel, *see* Cap

Pipe, inside diameter, 80; outside circumference, 80; outside diameter, 80; standard data, 80

Radial cutting, 2
Reducer, concentric, 71; eccentric, 73
Reducing lateral, 34
Reducing tee, branch enters header, 43; branch outside header, 48

Superimposing, 15

Torch, *see* Cutting
True Y, fitting main line template, 25, 27

81